Contractor's Guide to Green Building Construction

Contractor's Guide to Green Building Construction

Management, Project Delivery, Documentation, and Risk Reduction

Thomas E. Glavinich, D.E., P.E.

The University of Kansas

John Wiley & Sons, Inc.

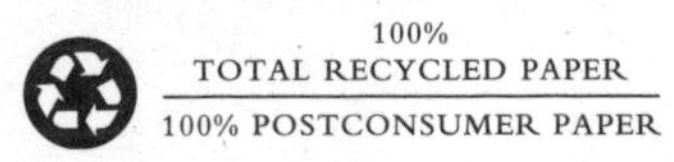

This book is printed on acid-free paper. ♾

Published by John Wiley & Sons, Inc., Hoboken, New Jersey
Published simultaneously in Canada

For general information about our other products and services, please contact our Customer Care Department within the United States at (800) 762-2974, outside the United States at (317) 572-3993 or fax (317) 572-4002.

Wiley also publishes its books in a variety of electronic formats. Some content that appears in print may not be available in electronic books. For more information about Wiley products, visit our web site at www.wiley.com.

Library of Congress Cataloging-in-Publication Data:

Glavinich, Thomas E.
The AGC contractor's guide to green building construction : management, project delivery, documentation, and risk reduction / Thomas E. Glavinich.
p. cm.
Includes bibliographical references and index.
ISBN 978-0-470-05621-9 (cloth)
1. Sustainable buildings—Design and construction. 2. Sustainable architecture. 3. Construction industry—Management. I. Title.
TH880.G53 2008
690—dc22

2007039330

Printed in the United States of America

10 9 8 7 6 5 4 3 2 1

Contents

Foreword

I remember when I was four or maybe five years old, seeing my grandfather pull up in my driveway in a great big dump truck. His father's name, his company's name, was painted in big red letters on the side of that bright white Mac truck. My grandfather owned a construction company, and on this sunny Saturday morning, he was delivering a load of sand for the sandbox in the backyard. This was my first realization that my grandfather and his father before him owned a construction firm. Seeing him standing out on the running board of that truck triggered a feeling inside of me—a feeling that he loved what he did and was proud to be a part of a great industry. As I grew older, that image never faded; it only grew stronger. By the time I was 10 years old, I knew that I wanted to work for my grandfather. I wanted to be a truck driver. As the years went by, I realized it was not a truck driver I wanted to be, it was being a part of the industry in which he was so successful.

Just like my direction in professions changed, so did the focus of this company. When my great-grandfather started his company, the primary focus was concrete flatwork. As the company grew, they moved into larger projects, and by the time I had that first glimpse into my legacy, the company had a primary focus in roads and bridges. By the time I was old enough to hold a real job, the company performed virtually no roadwork and had diversified into steel fabrication, commercial buildings, heavy industrial, healthcare, and the installation of process automation equipment. I progressed from a copy boy through the ranks of project management to the level of vice president before launching a subsidiary sustainability consulting firm. What does this evolution of one firm have to do with green building and this book?

In today's changing world, it is hard to read through a trade publication or a mainstream magazine without seeing a story about green buildings. Even magazines like *Vanity Fair* and *Sports Illustrated* have had issues that deal with this topic. When you read these articles, you see words used to describe the movement like *green, integrated design, sustainability, high-performance buildings, living buildings, integrated delivery, next-generation buildings*, and organizations like the Green Building Initiative (GBI), U.S. Green Building Council (USGBC), National Institute of Building Sciences (NIBS), General Services Administration (GSA), Environmental Protection Agency (EPA), and Architecture 2030 just to list a few. What does all of this mean? Who are all of these people, and why are they putting such a tremendous amount of effort into this movement when many of us may dismiss it as a passing fad?

When you look at the profile of most successful companies, you find that one key to success is the ability to change and adapt to the marketplace. The executives who run those successful companies scan the business horizon for indicators that show them which way to steer their firms. Those of us who are lucky enough to work in the construction industry are experiencing a change that has not been seen before. The external effects of global climate changes, diminishing amounts of raw and manufactured goods, a shrinking labor pool, an increased amount of governmental regulation, and rising energy costs combined with the internal forces of shorter time schedules, lower margins, a mix of delivery methods, and changes in the insurance market have forced all of us to reevaluate the way we deliver our services to the marketplace. We are seeing an evolutionary progression in the way buildings are designed and constructed. We are seeing the green movement take our industry by storm. All of us in the construction industry need to understand what this means and how it may affect our businesses.

In the recent past, many of us who work in the construction industry have been labeled as a "necessary evil," a part of the delivery supply chain that is bought and sold as a commodity, forced to compete using the lowest common denominator—price. Long past are the days when we were respected for the quality of services we delivered, when we were respected as "master builders." I feel that the green movement can be looked at in two ways. It can be viewed as just another trend that we as professionals need to understand in order to continue to compete using the lowest common denominator, but it could also be looked at as a way for the construction industry to regain a higher level of respect. Our ability to solve complex problems and demonstrate quality can help all of us regain the position of the master builder.

The Associated General Contractors of America (AGC) commissioned this book to aid construction professionals from any tier to understand what it takes to compete and to be more successful in the marketplace. The author and those who contributed to this book have drawn on extensive research and real-life experience to define the terms used in this segment of the industry, the risks associated with these types of construction projects, and how best to manage a "green" project. Our intent was to provide guidance to contractors. The structure of the book follows the progression of a project, from acquisition through project closeout. It delineates the typical project activities and explains how normal deliverables need to be modified when working on a green project. It is our hope that this book will help contractors navigate the challenges of green projects and emerge better able to take advantage of the multitude of benefits these projects offer.

THOMAS A. TAYLOR

General Manager
Vertegy, an Alberici Enterprise

Thomas Taylor is a member of the AGC Environmental Network Steering Committee and has served as Chair on the association's Green Construction Task Force. Thomas Taylor also participates on the ANSI Technical Committee charged with developing the first ever consensus based national standard on green building, Green Globes. The Alberici headquarters has received the highest level of certification provided both the U.S. Green Building Council and the Green Building Initiative programs, as well as other awards and recognition. Chapter 4 of this publication includes a case study on the Alberici headquarters.

Preface

Today, the state of the environment and our impact on it is a topic of discussion and debate at all levels of government, as well as in the corporate board room and across the table at the local diner. In the past, the focus was on the industrial and transportation sectors of our economy, and commercial and institutional buildings escaped a lot of public scrutiny because they appeared to be benign environmentally. Commercial and institutional buildings do not openly belch smoke, deplete natural resources, use foreign oil, or pollute the land or water supply. However, over the past decades we have become increasingly aware of the amount of energy being consumed by commercial and institutional buildings and the impact that their construction and operation has on our environment. Even though they do not openly appear to contribute to the environmental problems that we face today, their construction and operation results in all of the aforementioned environmental impacts as we clear land for their construction, use precious natural resources to manufacture the materials used to construct them, dispose of all sorts of waste throughout their life, and use fossil fuels directly in the form of natural gas or indirectly in the generation of electricity to operate them. The importance of the commercial sector and commercial and institutional buildings to our economy will grow in the future as the United States continues its shift from a manufacturing and industrial-based economy to a service economy, which will compound the problem if we do not take action today.

Economic growth is the key to improving our quality of life in the United States, and we will need to continue to build and operate commercial and institutional buildings in the future. We need these buildings to support our growing population and to provide healthy and productive environments for people to live, work, and play. Reconciling the seemingly opposing goals of improving the environment and providing the needed infrastructure to support economic growth is the genesis of the green building movement. The green building movement is simply about being mindful of the potential impact that the construction and operation of commercial and institutional buildings will have on the environment and devising innovative strategies to mitigate or eliminate these impacts. This means changing the way we think about buildings and their construction and operation as well as the integration of new state-of-the-art technologies into buildings. In short, we are finding that we can have our cake and eat it too. Today, building owners, designers, manufacturers, and contractors are focused on achieving increasingly environmentally friendly and energy- efficient buildings, with the ultimate goal of producing environmentally and energy-neutral buildings in the not-too-distant future. The green building movement will provide new opportunities and challenges for contractors as high-performance buildings move into the mainstream and become the norm in the United States.

Helping contractors recognize and address the challenges of green building construction is what this book is all about. Green buildings are as much about construction as they are about design. Green designs must be implemented and documented by the contractor for the building to become a reality and achieve its potential. Sustainable building design and construction can impact the contractor's material and equipment procurement, sequencing and scheduling of work, jobsite productivity, and commissioning and closeout activities. Green requirements may impat not only impact the direct costs of construction, but also project and home office overhead resulting from increased administrative, documentation, and commissioning requirements. All of these potential impacts must be recognized and accounted for when bidding, contracting for, constructing, and closing out a green building project.

The purpose of this book is to provide the contractor with a guide to effectively bidding, contracting for, constructing, and closing out a green building project. This book is focused on the green building process from the contractor's viewpoint and its potential impact on project management, delivery, documentation, and risk. This book is not about green design or the ins or outs of any particular green building rating system that have already been addressed in many other publications. Instead, this book is focused on the contractor's business and construction processes and how they may be affected by green construction. This book addresses the overall construction process, including understanding green project requirements for bidding, contracting for green construction, managing green design when the contractor is operating as a design-builder, subcontracting on green projects, green procurement, managing green construction, and green project commissioning and closeout. It is our hope that the information contained in this book will help contractors who are new to the green building market enter it more confidently and provide contractors who are already working in this market with new insights that will help them reduce their risk and be more competitive.

The green building market is an emerging market, and this book represents a snapshot of this new market today and contractors' experiences to date. The green building market will continue to evolve as more public and private building owners demand high-performance buildings and that their commitment to the environment be recognized through green building certification. Existing green building rating systems will continue to change, and new rating systems will emerge as sustainable buildings become increasingly mainstream, our understanding of what makes a green building continues to evolve, and new design and construction techniques, materials, equipment, and systems become available. Also, government agencies at all levels are beginning to require that public and private buildings under their jurisdiction be certified or certifiable using a specified third-party green building rating systems. This is an exciting time in the building industry, with many changes and opportunities for contractors. However, to be successful, contractors must keep up with the rapid pace of change in order to compete in this emerging market.

Thomas E. Glavinich, D.E., P.E.
The University of Kansas
Lawrence, Kansas

Acknowledgments

The author would like to thank The Associated General Contractors of America (AGC), the AGC staff, and the AGC membership for their help and support throughout this project. This includes the AGC Green Building Task Force members who freely gave of their time, knowledge, and experience. Without the AGC and the input from its members, this book would not have been possible.

I would also like to take this opportunity to recognize and thank several individuals who worked closely with me throughout this project, providing detailed reviews of chapter drafts and invaluable input and suggestions. These include Melinda Tomaino, Associate Director of Environmental Services at AGC; Thomas Taylor, Vertegy: An Alberici Enterprise; Joshua Bomstein, Creative Contractors, Inc.; and Daniel Osterman, McGough Construction Company, Inc.

AGC GREEN BUILDING TASK FORCE

Joshua Bomstein	Creative Contractors, Inc.	Clearwater, Florida
Dirk Elsperman	Tarlton Corporation	St. Louis, Missouri
Chris Miller	Brasfield & Gorrie, LLC	Birmingham, Alabama
Daniel Osterman	McGough Construction Company, Inc.	St.Paul, Minnesota
Kimberly Pexton	HITT Contracting, Inc.	Fairfax, Virginia
Beth Studley	Holder Construction Company	Atlanta, Georgia
Thomas Taylor	Vertegy: An Alberici Enterprise	St. Louis, Missouri
Melinda Tomaino	The Associated General Contractors	Arlington, Virginia
Mark Winslow	Gilbane, Inc.	Providence, Rhode Island

chapter 1

Green Construction and the Contractor

1.1 INTRODUCTION

Concern about the environment and the future of our planet has become the focal point of everyday conversation, political debate, and media coverage in the United States today. The United States currently uses a significant amount of the world's energy and produces a significant portion of the world's greenhouse gases. This debate was focused on the industrial, manufacturing, and transportation sectors in the past, but energy usage and its associated environmental impacts have become a major issue in the building industry. Commercial and residential buildings consume about 40 percent of the energy used in the United States, according to the U.S. Department of Energy's Energy Information Agency [EIA 2007]. In addition, both the amount of energy used in buildings and its percentage of the total U.S. annual energy usage is expected to increase in the coming decades despite conservation initiatives, increased building efficiency, and rising energy costs. As a result, more building owners, including all levels of government, are demanding high-performance buildings and are seeking third-party certification to verify and publicly recognize their commitment to the environment.

All of this change has put the construction industry in a reactive mode as it adjusts to the new technical and administrative requirements that are being imposed by the project contract documents and third-party certification requirements. However, green construction doesn't have to be just another contract requirement and associated risk that the contractor must address.

Instead, the contractor can embrace the principles of green construction and become proactive, which is not only good for the environment but also good for business. This chapter introduces green construction and the role of the contractor in creating a sustainable environment through green construction practices.

1.2 SUSTAINABLE DEVELOPMENT: WHAT IS IT?

The World Commission on Environment and Development (WCED) developed a definition of *sustainable development* that was included in its 1987 report. This report has become known as the Brundtland Report after the chair of the commission, G. H. Brundtland. The WCED's definition has been widely accepted since its publication and is as follows [WCED 1987]:

> Humanity has the ability to make development sustainable—to ensure that it meets the needs of the present without compromising the ability of future generations to meet their own needs. The concept of sustainable development does imply limits—not absolute limits but limitations imposed by the present state of technology and social organization on environmental resources and by the ability of the biosphere to absorb the effects of human activity.

The key phrase in the WCED's definition of *sustainable development* is that sustainable development "meets the needs of the present without compromising the ability of future generations to meet their needs." Buildings constructed and renovated today should have a useful life of 30 or more years. Construction plays an important role in sustainable development because it uses the earth's resources to build the buildings where people live, work, and play. Just like every other industry, the construction industry is responsible for the environment. Contractors can strive to ensure that the construction process is efficient, uses renewable resources, and minimizes resource use and waste within the confines of the owner's construction procurement process and contract documents.

1.3 GREEN BUILDING DEFINED

The term *green building* is defined in the American Society of Testing and Materials (ASTM) Standard E2114-06a as a building that provides the specified building performance requirements while minimizing disturbance to and improving the functioning of local, regional, and global ecosystems both during and after its construction and specified service life [ASTM E2114]. This definition illustrates the importance of the construction process in the making of a green building. During construction, it is about minimizing the environmental impact of the construction process on the environment through procurement, site layout and use, energy use, waste management, and construction operations. However, the contractor's impact on building sustainability doesn't stop at substantial completion. A project delivery system that involves the contractor in the design process or provides leeway in the contract documents may allow the contractor to use materials and installation techniques based on its expertise and experience that will minimize operation and maintenance (O&M) costs over the life of the building, provide a more durable facility, reduce building-related illnesses that impact the well-being and productivity of building occupants, and maximize the reuse of building materials at the end of the building's life.

1.4 GREEN CONSTRUCTION: REACTIVE OR PROACTIVE?

From a reactive standpoint, green construction only occurs because of the requirements contained in the contract documents. The contractor builds the building in accordance with the project plans and specifications and is only passively involved in sustainable development. This is mainly how the construction industry has approached green construction in the past. However, it is becoming more difficult for the contractor to sit on the sidelines and not be actively involved in green construction. In addition to being socially responsible and good corporate citizens, contractors should become involved in green construction for several reasons, including the following:

- Owners are demanding that their suppliers demonstrate their commitment to the environment and provide environmentally sustainable products and services. This includes the construction services provided by contractors and may offer a competitive advantage for environmentally proactive contractors.
- Many contractors' field and office employees are concerned about the environment and prefer to work for an environmentally conscious firm and feel that they are contributing to the solution rather than being part of the problem.

- Environmental laws and regulations that contractors are subject to are increasing, along with the liability, fines, and cleanup costs associated with noncompliance.
- Overhead costs associated with complying with environmental laws and regulations, as well as insurance premiums for environmental coverage, are increasing.
- The public is becoming increasingly concerned about the environment, and there is increasing focus on all industries including construction that could result in increased governmental regulation and compliance costs for those industries that do not take a proactive approach to the environment.

In order to take a proactive approach to green construction, contractors need to make the environment a key element in both their business strategy and day-to-day operations.

1.5 GREEN CONSTRUCTION WITHOUT GREEN DESIGN?

The owner decides if it is going to build green and the extent to which the project will be sustainable during the early planning stages of the project. The design team then implements the owner's decision during the design process. Whether or not the contractor is involved in the design process, its primary function in the project delivery process is to convert the design team's design into physical reality for the owner for the agreed-upon price and within the agreed-upon time frame.

The contractor's expertise is in planning and managing the construction process and procuring the necessary labor, materials, and equipment to get the job done by either self-performing the work or contracting specific portions of the work to specialty contractors. As a result of its expertise, the contractor has sole control of the construction means, methods, techniques, sequences, and procedures unless limited by specific contract provisions in order to ensure both safe and efficient project delivery. With or without a green design, the contractor has control of the construction process and can take a proactive stance with respect to the environment and green construction. Building green does not have to be just another contract requirement that needs to be addressed during the building construction and commissioning process. The contractor can build green with or without a green design.

1.6 WHAT IS GREEN CONSTRUCTION?

Green construction is defined in this book as follows:

> Green construction is planning and managing a construction project in accordance with the contract documents in order to minimize the impact of the construction process on the environment.

This definition places the contractor in a proactive position with regard to the environment. The contractor bids or negotiates the work in accordance with the contract documents, as it always has, being mindful of selection criteria that the owner will use to select a contractor for the project. Then in planning and managing the work, the contractor's project team looks for ways to minimize the impact of the construction process on the environment, which includes (1) improving the efficiency of the construction process, (2) conserving energy, water, and other resources during construction, and (3) minimizing the amount of construction waste, among other strategies that do not adversely impact its project budget or schedule and may even reduce costs and increase productivity. Being green can be a winning proposition for the contractor.

1.7 GREEN IS LEAN

Lean construction is all about removing waste from the contractor's business and construction processes in order to make it more efficient. Green construction is also focused on removing waste from the construction process and adds an environmental dimension to lean construction. On a renovation or demolition project, the deconstruction of the existing facility can yield a significant amount of material that can be either recycled or reused, possibly creating a profit center for the contractor, diverting waste from landfills, and conserving energy and resources through recycling. When procuring materials and equipment, the contractor can work with suppliers to better package and bundle materials that could both reduce waste and improve productivity. Similarly, off-site prefabrication of materials can reduce waste at the jobsite and improve productivity. When expendables such as sealants and adhesives are purchased, low-emitting materials that can improve the working environment and productivity could be substituted for traditional materials.

1.8 THE GREEN CONTRACTOR

Being green needs to become a way of doing business and part of the contractor's corporate culture. Within the home office, the contractor should investigate ways that will promote and demonstrate its commitment to the environment as well as provide a payback when possible. This could include anything from the use of recycled copy paper to photovoltaics. The existing lighting system could be retrofitted with energy-efficient lamps and ballasts, as well as occupancy sensors and daylighting controls where appropriate. As office equipment and appliances are replaced, the new equipment and appliances could be certified as energy-efficient. Even replacement company vehicles could be hybrid vehicles or run on alternate fuels.

1.9 ADVANTAGES OF BEING GREEN

Improved productivity and reduced costs at the jobsite, as well as reduced home office overhead costs, provide tangible benefits of going green that the contractor can measure. However, other advantages of being green may be more difficult to quantify but may benefit the contractor. For one, by focusing on green construction every day, the contractor's personnel will become more knowledgeable about the possibilities. So that when an actual green construction project comes along, they will have a better understanding of the actual work and costs involved and be more effective construction team members. Similarly, a commitment to being green in both the office and the field will appeal to many of the contractor's employees. Enlisting their help in this initiative will build camaraderie and commitment to the firm. Finally, the contractor's clients are becoming increasingly environmentally conscious and are looking for the same commitment in the firms with which they work.

1.10 OVERVIEW OF THIS BOOK

This book provides a blueprint for becoming a green contractor. This chapter has started the process by defining *green construction* in such a way that the contractor can take a proactive approach to sustainability that should benefit itself, its employees and other stakeholders, and society as a whole. Chapter 2 addresses the elements of green construction that impact both home office and field operations. It also discusses the various green building standards and rating systems to provide the contractor with an overview of the systems and their requirements.

The key to reducing risk associated with green construction is to thoroughly understand the project requirements. Bidding green construction is covered in Chapter 3. This chapter builds on previous chapters by pointing out where in the construction documents the contractor can expect to find green construction requirements and their possible impact on cost, schedule, and productivity throughout the construction process from mobilization through commissioning. Green construction requirements that address technical and administrative requirements are cross-referenced with both the 1995 and 2004 editions of the Construction Specifications Institute's (CSI) *MasterFormat*™ that are currently in use.

Contracting for green construction and the construction contract's importance in managing the contractor's risk are addressed in Chapter 4. This chapter opens with a discussion of contractual risk and risk management on construction projects and then discusses how green construction requirements can impact the contractor's role and responsibilities under common project delivery systems, including design-bid-build, construction manager agency and at risk, and design build. Typical contract documents and contract requirements that could impact the contractor on a green construction contract are then covered. This chapter closes with a discussion of insurance coverage and bonding requirements on green projects.

The use of design build as a project delivery system by owners is growing in the United States for commercial and institutional facilities. As a result, the contractor that is assuming the role of design builder on a green project needs to understand how to manage not only construction but also the design process. Chapter 5 addresses managing green design on design-build projects. This chapter starts with obtaining the services of a qualified designer for a green project and the role of the designer during construction and commissioning. Topics addressed in this chapter include defining design services needed, soliciting proposals and selecting a designer, contracting with the designer, professional liability insurance, overseeing the design process, using design reviews to ensure that the owner's green requirements are being met within budget and on schedule, and project closeout. This chapter should also be useful to the construction manager in working with the owner and designer during the project planning and design process.

Subcontracting portions of the project work to specialty contractors on green construction projects results in some unique challenges for the contractor. The success of a green construction project depends on subcontractor performance, which means that specialty contractors must understand their roles and responsibilities. Chapter 6 focuses on the unique challenges faced by the contractor subcontracting work on a green project. This chapter covers subcontractor qualifications and the need to prequalify subcontractors for green projects, defining subcontract scope of work on green projects, educating subcontractors about their responsibilities, training subcontractors to fulfill their responsibilities, green subcontract terms and conditions, involving subcontractors in the planning and scheduling process, building system commissioning, and project closeout.

Materials and equipment are a critical factor in green building construction and the major portion of criteria used to classify or certify a green building. Even though the designer specifies materials and equipment, the contractor and its subcontractors must understand the material and equipment specifications and the characteristics that make the materials and equipment green. This makes material and equipment procurement a critical success factor for the contractor in any green construction project. Chapter 7 addresses procuring material and equipment for green building projects, with a focus on understanding green material criteria and terminology. This chapter will not only help the contractor procure the right materials and equipment for the work it self-performs, but will also help it qualify material and equipment suppliers, review subcontractor material and equipment submittals for contract compliance, provide more effective value analyses and constructability reviews, and better meet project closeout material and equipment documentation and certification requirements.

Chapter 8 addresses constructing a green project and those aspects of sustainable construction that specifically impact the contractor's construction operations. Topics that are covered include using site layout to minimize site disturbance, erosion, and runoff during construction; minimizing the use of fossil fuel and emissions through conservation and alternate fuels; reducing waste through material recycling and reuse; and improving indoor air quality during construction by using low-emitting materials, among other strategies. Also covered in this chapter are the measurement and documentation requirements that may be imposed on the contractor during construction by the contract documents or third-party green building certification process. This chapter should also be helpful to the contracting firm that wants to be more proactive environmentally and incorporate green building methods into its day-to-day construction operations.

Commissioning and closeout of a green construction project is more complex than a traditional building project, particularly if the owner is seeking third-party certification. Chapter 9 addresses green building commissioning, including understanding the contractual requirements for commissioning, the need for a comprehensive mutually agreed-upon commissioning plan early in the project, working with an outside owner-appointed commissioning agent, typical requirements for system startup and testing, and typical documentation that needs to be submitted. In addition, typical contract closeout requirements for green buildings are also addressed in this chapter, including submitting project documentation such as record drawings, addressing warranties and guarantees required by the contract documents, and training the owner's personnel.

At the end of each chapter short case studies have been included to illustrate the green building topics covered in that chapter. These case studies were provided by AGC members involved in green building construction.

Appendix A provides a glossary of terms and abbreviations used in the green building industry and throughout this book. Appendix B provides a list of references and additional resources for the contractor.

1.11 CASE STUDIES

The Associated General Contractors of America New Headquarters

The Navy League Building

In 2005, AGC moved its new headquarters offices into the Navy League Building at 2300 WIlson Boulevard in Arlington, Va. The building exemplifies "green building" advances and received a high "Silver Rating" under the Leadership in Energy & Environmental Design (LEED®) Green Building Rating System.

Site

The site meets quality growth principles, which include sensitivity to the following criteria: location, building density, design, diversity, transportation, accessibility, environment and community. This redevelopment project is "transit-friendly," located one block from the Arlington Courthouse Metro station and on multiple bus routes. The building incorporates other transportation alternatives: bicycle storage and changing rooms for the building's occupants, charging stations for electric automobiles within the four-tier underground parking deck, and preferred parking for car and vanpools. To encourage the utilization of transportation alternatives, parking capacity meets only the minimum local zoning standards.

Figure 1-1 Photo courtesy of The Navy League.

Water

The building has a very advanced water efficiency system that aims to minimize the amount of potable water consumed by the project while simultaneously reducing the amount of storm water runoff from the site. A storm water detainment system catches rain-water and stores it in a large vault at the basement level so that it can be used for irrigating trees and shrubs on the property as well as for flushing the building's toilets. Low-flow fixtures, dual-flush toilets, and using recovered water for chiller re-supply will cut water use by over 30 percent, compared to conventional office buildings. The combination of the storm water reuse system and the high water efficiency plumbing fixtures allows the building to use approximately 60 percent less potable water overall.

Energy

Building HVAC systems will increase heating and cooling efficiency, reducing operating costs and air pollution. Additionally the heating, cooling and refrigeration systems will not use any ozone depleting CFCs or HCFCs. An Energy-Star Rating roofing system will decrease temperatures at the roof level. This roof system helps reduce both the radant heat load of the building and lowers temperature at the roof thereby alleviating the impact of the building on urban heat islands. Exterior lighting will be designed to minimize light pollution and assist national dark sky initiatives. Overall building energy use will be approximately 20 percent less than conventional office buildings.

Green Materials

The contractor—James G. Davis Construction—was required to salvage or recycle 75 percent of the waste from demolition, construction and land clearing. Twenty percent of the building materials had to come from within a radius of 500 miles, and at least 50 percent of the wood-based materials had to be Forest Stewardship Council certified.

Indoor Air Quality

Low-emission adhesives, paints and carpets along with exhaust systems designed to remove airborne particulate matter will improve indoor air quality.

R. J. Griffin & Company

Southface Eco Office, Georgia

When Southface Energy Institute, an Atlanta-based non-profit organization promoting sustainable homes, workplaces, and communities through education, research, advocacy and technical assistance, needed more office space, it decided to showcase sustainable strategies, materials, and products for commercial construction in its new building.

A unique and nontraditional project delivery method was used on the Eco Office, a three-story, 10,000-square-foot showcase of integrated design and 'state-of-the-shelf' green building products tracking LEED Platinum certification in Fall 2007. When Southface set out to expand its facility as an environmentally-responsible commercial demonstration project, it enlisted the design services of Lord Aeck & Sargent and the construction management services of a Green Building Consortium—five general contractors with LEED experience—DPR, Hardin, Holder, Skanska, and Winter, later assisted by RJ Griffin. This collective contribution, along with dozens of donated products and services, resulted in a model eco-conscious building at half the market cost.

Figure 1-2 Photo courtesy of Lord Aeck & Sargent.

Basing a building program on donated components, however, brought unforeseen challenges to the owner as well as the design and construction teams. As word of the project spread, an increasing number of generous manufacturers and subcontractors offered their products and services, relieving budget constraints, but often impacting design decisions and coordination.

No compromises were made in resource efficiency, however. Scheduled for completion in Fall 2007, the Eco Office is designed to use 60 percent less energy and 80 percent less potable water than a conventional office by incorporating a high-efficiency thermal envelope, daylighting strategies, efficient fixtures, salvaged photovoltaic panels, an extensive green roof and rainwater harvesting to eliminate potable water use for sewage conveyance, the innovative mechanical systems and irrigation.

The Eco Office collaboration has created a valuable exchange of information and ideas on sustainable design, building science, and product development that has strengthened the entire team and influenced the marketplace as a whole.

The best lesson learned by RJ Griffin, besides the fact that we enjoyed working with our competitors toward a worthwhile goal, is that, just like all commercial projects, budget and other obstacles can be overcome with a creative team of all members (owner, design team, and contractors) when everyone is involved early on in the preconstruction process. Green building just adds one more component to the challenge!

1.12 REFERENCES

ASTM International, Standard Terminology for Sustainability Relative to the Performance of Buildings, ASTM Standard E 2114–06a, 2006.

Energy Information Agency, www.eia.doe.com, 2007.

World Commission on Environmental Development, *Our Common Future*, New York, Oxford Press, 1987, p. 43.

chapter 2

Elements of Green Construction

2.1 INTRODUCTION

Green buildings are moving into the mainstream of the U.S. construction industry. Increasingly, private and public owners are requiring that their building projects be designed and constructed in an environmentally responsible manner and be recognized as green buildings. This is usually accomplished by requiring that the project achieve certification as a green building using a third-party rating system. Green building rating systems are typically point based and require that the project earn a certain number of points or percentage of applicable points in order to be certified to a given level. The various criteria that must be met to earn these certification points cover everything from site selection to incorporating renewable energy sources such as photovoltaics into the project. These requirements impact not only the building design but also the building construction. The contractor needs to be aware of these green rating systems and their requirements, because they can impact both the contractor's scope of work and costs.

This chapter provides a brief introduction to green rating systems and green building certifications in the United States. There are a number of regional, national, and international green building rating systems in use today both around the world and throughout the United States. As noted in Chapter 1, the green building market is new and evolving rapidly. As a result, existing green building rating systems are continually changing and evolving to address new knowledge and technology, meet the needs of an expanding market for green buildings, and incorporate state-of-the-art design, construction, and operation techniques.

In addition, existing green building rating systems are being adapted and new green building rating systems are being developed by various organizations to better fit their needs by addressing specific building types, building functions, or perceived shortcomings in the existing rating systems. Therefore, this chapter should be viewed as a guide to understanding green rating systems and the elements of green buildings as reflected in the rating systems rather than as a tutorial covering the specific technical and administrative requirements of any system. For a particular project, the contractor should look to the specific green building rating system being used on that project to understand its role and responsibilities for that project. To assist the contractor in finding the latest information about a particular green building rating system and certification requirements, contact information for the sponsoring organizations has been included in this chapter, and references are included in Appendix B.

2.2 GREEN BUILDING EVALUATION SYSTEMS

Several green building evaluation systems are in use today. These systems are being developed and promulgated at the international, national, regional, and local levels. All of these systems are similar in the green building criteria that they address, but they can be very different in their intent, criteria, emphasis, implementation, and other important ways. These differences are often a result of the goals of the sponsoring organization as well as the specific niche or void in the construction market they are intended to address. These differences in criteria really stand out when comparing the requirements of green building evaluation systems developed for different building types, regions of the country, or points in a building's life cycle. Today, you can find specific green building requirements tailored to public and private building owners and operators, building types such as schools and office buildings, locations such as state or city, and points in the building's life cycle such as new construction and operation.

All of these evaluation systems can be used to guide the design and construction of green buildings. The goal of green building evaluation systems is usually to ensure that the constructed building meets the owner's operational requirements while minimizing the impact of the building on the environment throughout its life, providing a healthy and productive environment for occupants, and reducing building energy usage and operating costs. These evaluation systems are typically designed to promote building systems' integration and the optimization of the building as a whole rather than optimizing individual stand-alone systems. This is achieved by encouraging a thorough planning process, integrated building design and construction, rigorous building commissioning and closeout procedures, and ongoing monitoring and evaluation of postoccupancy building performance.

Some of these green building evaluation systems also include a third-party evaluation process that leads to certification of the building as a green building. These third-party certification processes are intended to provide a quantitative method for measuring a building's greenness and an objective method for designating a building as a green building. Most of these green building evaluation systems are point based and certify that a building is green based on the number of points achieved or a percentage of points possible. Beyond certification as a green building, many of these third-party rating systems also grade the greenness of the building by awarding higher grades of certification for higher numbers or percentages of points.

In general, green building evaluation systems are voluntary. The exception to this is when the federal, state, or local government requires that public or private buildings within its jurisdiction be certified or certifiable at a specified level using one of these green building evaluation systems. If there is no outside requirement, then the owner decides whether it wants to use one of the green building evaluation systems on its project and, if so, the criteria on which to base the design and construction. As noted previously, each of these green building evaluation systems is different, and the owner needs to select the evaluation system that best meets its sustainable goals for the building project. The contractor also needs to understand that the green evaluation system selected by the owner will probably impact its scope of work and costs when compared to the same building project using traditional criteria or another green building evaluation system.

2.3 LEED™ CERTIFICATION

2.3.1 U.S. Green Building Council

The U.S. Green Building Council (USGBC) is an industry organization whose membership consists of all parts of the construction industry, including owners, designers, and contractors. The USGBC promotes the construction of environmentally friendly, high-performance buildings through its sponsorship of the Leadership in Energy and Environmental Design (LEED™) green building rating systems. The purpose of these rating systems is to provide an objective standard for certifying that a building is environmentally friendly or green. As a result of the public's growing concern about the environment and rising energy costs, there is a growing movement among both public and private building owners to have their buildings LEED™ certified. Although the foundation for LEED™ certification is laid during the design process, the design intent must be implemented through the construction process. The contractor needs to be aware of LEED™ requirements, because they can impact material and equipment procurement as well as construction requirements and costs, which will be discussed in the chapters that follow. In addition, understanding LEED™ requirements will allow the contractor to effectively analyze and value-engineer the project within the LEED™ requirements for the owner.

2.3.2 LEED™ Rating Systems

The USGBC LEED™ rating system started with new construction and major renovations (LEED™-NC) in 1999. Since that time, the USGBC has either developed or is in the process of developing rating systems to address the specific needs and characteristics of other building types and projects. Currently, LEED™ rating systems include those shown in the following table:

Table 2-1

		Version	
LEED™ Designator	**Rating System Purpose**	**No.**	**Date**
LEED-NC	New Construction and Major Renovations	2.2	2005
LEED-CS	Core and Shell	2.0	2006
LEED-CI	Commercial Interiors	2.0	2005
LEED-EB	Existing Buildings: Upgrades, Operations, and Maintenance	2.0	2005
	Homes (Pilot)	1.11a	2007
	Neighborhood Developments (Pilot)		2007
	Retail: New Construction and Major Renovation (Pilot)	2.0	2007
	Schools: New Construction and Major Renovation (Pilot)		2007
	Multiple Buildings and On-Campus Building Projects (Pilot)		2007

LEEDTM-NC was the first USGBC LEEDTM rating system, and it is currently the rating system in the widest use throughout the United States. Therefore, the following sections will focus on LEEDTM-NC, because this is the most likely LEEDTM rating system that the contractor will encounter. The LEEDTM rating systems for core and shell (LEEDTM-CS), commercial interiors (LEEDTM-CI), and existing buildings (LEEDTM-EB) are similar and based on LEEDTM-NC.

2.3.3 LEEDTM-NC Certification Process

LEEDTM-NC certification of a building project starts with the owner's decision that the project will be a green project. In the early stages of design, the owner registers its intent to have the building project LEEDTM-NC certified with the USGBC. The owner's decision and registration must happen early in the design process, because the decision to have a building LEEDTM-NC certified will drive many fundamental decisions throughout the design process and could even impact site selection if the project site has not already been determined.

As part of the registration process, the owner establishes goals for the project in the following six categories in conjunction with the contractor and/or architect as shown in Figure 2-1:

- Sustainable Site (SS)
- Water Efficiency (WE)
- Energy and Atmosphere (EA)
- Materials and Resources (MR)
- Indoor Environmental Quality (EQ)
- Innovation and Design Process (ID)

LEEDTM-NC certification is based on the owner's ability to demonstrate that the building project meets the requirements of the LEEDTM-NC rating system. The LEEDTM-NC rating system is summarized in Figure 2-1 and includes the six categories listed plus specific subcategories designated either as prerequisites or credits under each category. As can be seen from Figure 2-1, prerequisites do not have any points associated with them but must be achieved in order for the building to earn any points for that category. For instance, the Material and Resources (MR) category has one prerequisite associated with it. In order for the project team to earn any credits toward

CREDIT		DESCRIPTION	LEED PTS
		SUSTAINABLE SITES	
SS	P 1	Construction Activity Pollution Prevention	
SS	C 1	Site Selection	1
SS	C 2	Development Density & Community Connectivity	1
SS	C 3	Brownfield Redevelopment	1
SS	C 4.1	Alternative Transportation: Public Transportation Access	1
SS	C 4.2	Alternative Transportation: Bicycle Storage & Chancing Rooms	1
SS	C 4.3	Alternative Transportation: Low Emitting & Fuel Efficient Vehicles	1
SS	C 4.4	Alternative Transportation: Parking Capacity	1
SS	C 5.1	Site Development: Protect Or Restore Habitat	1
SS	C 5.2	Site Development: Maximize Open Space	1
SS	C 6.1	Storm Water Design: Quantity Control	1
SS	C 6.2	Storm Water Design: Quality Control	1
SS	C 7.1	Heat Island Effect: Non-Roof	1
SS	C 7.2	Heat Island Effect: Roof	1
SS	C 8	Light Pollution Reduction	1
		WATER EFFICIENCY	
WE	C 1.1	Water Efficient Landscaping: Reduce By 50%	1
WE	C 1.2	Water Efficient Landscaping: No Potable Water Use Or No Irrigation	1
WE	C 2	Innovative Wastewater Technologies	1
WE	C 3.1	Water Use Reduction: 20% Reduction	1
WE	C 3.2	Water Use Reduction: 30% Reduction	1
		ENERGY & ATMOSPHERE	
EA	P 1	Fundamental Commissioning Of The Building & Energy Systems	
EA	P 2	Minimum Energy Performance	
EA	P 3	Fundamental Refrigerant Management	
EA	C 1	Optimize Energy Performance	1–10
EA	C 2	On-site Renewable Energy	1–3
EA	C 3	Enhanced Commissioning	1
EA	C 4	Enhanced Refrigerant Management	1
EA	C 5	Measurement & Verification	1
EA	C 6	Green Power	1
		MATERIALS & RESOURCES	
MR	P 1	Storage & Collection Of Recyclables	
MR	C 1.1	Building Reuse: Maintain 75% Of Existing Walls, Floors, & Roof	1
MR	C 1.2	Building Reuse: Maintain 95% Of Existing Walls, Floors, & Roof	1

Figure 2-1 LEED™-NC Prerequisites and Credits.

CREDIT		DESCRIPTION	LEED PTS
MR	C 1.3	Building Reuse: Maintain 50% Of Interior Non-Structural Elements	1
MR	C 2.1	Construction Waste Management: Divert 50% From Disposal	1
MR	C 2.2	Construction Waste Management: Divert 75% From Disposal	1
MR	C 3.1	Material Reuse: 5%	1
MR	C 3.2	Material Reuse: 10%	1
MR	C 4.1	Recycled Content: 10% (Post-Consumer + 1/2 Pre-Consumer)	1
MR	C 4.2	Recycled Content: 20% (Post-Consumer + 1/2 Pre-Consumer)	1
MR	C 5.1	Regional Materials: 10% Extracted, Processed & Manufactured Regionally	1
MR	C 5.2	Regional Materials: 20% Extracted, Processed & Manufactured Regionally	1
MR	C 6	Rapidly Renewable Materials	1
MR	C 7	Certified Wood	1
		INDOOR ENVIRONMENTAL QUALITY	
EQ	P 1	Minimum IAQ Performance	
EQ	P 2	Environmental Tobacco Smoke (ETS) Control	
EQ	C 1	Outdoor Air Delivery Monitoring	1
EQ	C 2	Increased Ventilation	1
EQ	C 3.1	Construction IAQ Management Plan: During Construction	1
EQ	C 3.2	Construction IAQ Management Plan: Before Occupancy	1
EQ	C 4.1	Low-Emitting Materials: Adhesives & Sealants	1
EQ	C 4.2	Low-Emitting Materials: Paints & Coatings	1
EQ	C 4.3	Low-Emitting Materials: Carpet Systems	1
EQ	C 4.4	Low-Emitting Materials: Composite Wood & Agrifiber Products	1
EQ	C 5	Indoor Chemical & Pollutant Source Control	1
EQ	C 6.1	Controllability Of Systems: Lighting	1
EQ	C 6.2	Controllability Of Systems: Thermal Comfort	1
EQ	C 7.1	Thermal Comfort: Design	1
EQ	C 7.2	Thermal Comfort: Verification	1
EQ	C 8.1	Daylight & Views: Daylight 75% Of Spaces	1
EQ	C 8.2	Daylight & Views: Daylight 90% Of Spaces	1
		INNOVATIVE DESIGN	
ID	C 1.1	Innovation In Design	1
ID	C 1.2	Innovation In Design	1
ID	C 1.3	Innovation In Design	1
ID	C 1.4	Innovation In Design	1
ID	C 2	LEED Accredited Professional	1
TOTAL POINTS POSSIBLE & REQUIRED PROJECT POINTS			**69**

Figure 2-1 (*Continued*)

LEED™-NC certification in the Material and Resources category, the requirements associated with prerequisite P1, which addresses the storage and collection of recyclables, must be met.

Once a category's prerequisites are met, points toward LEED™-NC certification can be achieved by meeting the requirements of the various credits that are included as part of the category. As can be seen from Figure 2-1, many credits are broken down so that additional points can be awarded based on the level of achievement. For example, in the Indoor Environmental Quality (EQ) category, Credit 8 addresses daylight and views. If the building does not meet the criteria of Credit 8.1, which is daylighting in 75 percent of the building spaces, then no points toward LEED™-NC certification will be awarded for Credit 8. However, if the requirements for Credit 8.1 are met and 75 to 89 percent of spaces are daylit, then one point is earned by the project toward LEED™-NC certification. Similarly, if more than 90 percent of the building spaces meet the daylighting criteria set forth in Credit 8.2, then two points will be earned toward LEED™-NC certification.

Throughout design and construction, the project team documents how they are meeting both the category prerequisites and credits for points toward certification as shown in Table 2-2. Beyond fulfilling category prerequisites, the owner is free to determine what categories and credits within those categories that will be sought to obtain certification. Not every credit within the LEED™ rating system needs to be addressed in the building design and construction. The number of credits earned by the project will, however, determine the level of LEED™ certification. Credits required to earn the various levels of LEED™ certification are as follows:

There are multiple submittals during the LEED™ application process, including some preliminary design submittals. Whoever submitted the registration form (owner, contractor, or architect) would also be responsible for submitting the application to the USGBC for LEED™ certification. This would consist of the required project documentation to substantiate each prerequisite and credit claimed, other supporting documentation including

Table 2-2

Certification Level	Points Required
Certified	26–32
Silver	33–38
Gold	39–51
Platinum	52–69

a project narrative that includes at least three project highlights, and the application fee. Following receipt and review of the application, the USGBC issues its preliminary findings, along with a request for any additional information that it needs to perform its final review. Within 30 days, the project team makes its final submittal to the USGBC, which is followed by the USGBC's final review and award of LEED™ certification to the project.

2.3.4 Information on the LEED™ Rating System

To assist the project team in meeting the requirements of the LEED™-NC rating system, the USGBC publishes the *LEED™ Reference Guide* and *LEED™ Letter Templates*. The *LEED™ Reference Guide* provides invaluable information about the intent of each prerequisite and credit, requirements and submittals needed, design strategies, case studies, and other information. *LEED™ Letter Templates* help the construction team prepare the LEED™ certification application by providing electronic forms for documenting that prerequisite and credit performance requirements have been met. Both the LEED™ rating system and letter templates are available on the USGBC Web site, and the reference guide can be purchased from the USGBC. The USGBC's contact information is as follows:

U.S. Green Building Council
1800 Massachusetts Avenue NW, Suite 300
Washington, D.C. 20036
Telephone: (202) 828-7422
Fax: (202) 828-5110
E-mail: info@usgbc.org
Web site: www.usgbc.org

2.3.5 Projects, Not Products, Are LEED™ Certified

Only building projects can be LEED™ certified. The USGBC does not certify building products for use on LEED™ projects. The focus of LEED™ prerequisites and credits is on building life-cycle performance and meeting specific goals that are aimed at improving the environment, reducing energy use, and increasing occupant comfort and productivity. The LEED™ rating system is really about optimizing the building as a system rather than optimizing individual systems or components, which may lead to less optimal building performance as a whole.

2.3.6 LEED™-Accredited Professionals

While not required, a LEED™-Accredited Professional can be a valuable asset to a building project seeking accreditation because of his or her knowledge of the process and requirements. Anyone can become a LEED™-Accredited Professional by taking the USGBC's examination that tests the candidate's knowledge of green construction, sustainable building design, and the LEED™ certification process. Having a LEED™-Accredited Professional on a building project is worth one extra point toward accreditation under Credit 2 of the LEED™-NC Innovative Design (ID) Category.

Having one or more LEED™-Accredited Professionals on staff may be advantageous for the contractor both from an operational and marketing standpoint. Operationally, someone familiar with the LEED™ accreditation requirements and process would be valuable in value engineering, bid preparation, procurement, and project closeout where specific documentation is required. For instance, in the Materials and Resources (MR) Category, there are prerequisites and credits for construction waste management, use of recycled building materials, and use of materials manufactured within a 500-mile radius. These and other LEED™ prerequisites and credits could impact the contractor's project costs and productivity. In addition, having a LEED™-Accredited Professional could also be a valuable marketing tool, because it shows that the contractor is interested in sustainable construction and understands LEED™ requirements, giving it a competitive advantage.

2.4 GREEN GLOBES™ CERTIFICATION

2.4.1 Origins of Green Globes™

The Building Research Establishment Environmental Assessment Method (BREEAM) was developed in the United Kingdom as a voluntary green building rating system. The purpose of BREEAM is to assess the energy and environmental performance of existing commercial buildings along the following four dimensions:

- Energy Efficiency
- Environmental Impact

- Health
- Operation and Management

BREEAM was adopted by the Canadian Standards Association (CSA) and published as BREEAM Canada for Existing Buildings in 1996. BREEAM Canada is administered by ECD Energy and Environmental, Ltd.

The BREEAM Green Leaf program evolved along with Green Leaf for Municipal Buildings by the Federation of Canadian Municipalities. These programs evolved into BREEAM Green Leaf for the Design of New Buildings in 2000. The objective of BREEAM Green Leaf was to provide building owners with a self-assessment tool that can be used to evaluate existing buildings. ECD developed a Web-based version of Green Leaf and dubbed it Green Globes in 2004.

2.4.2 Green Building Initiative

The Green Building Initiative (GBI) was established in 2004 as a nonprofit organization to help local homebuilding associations promote the use of the National Association of Homebuilders' (NAHB) *Model Green Home Building Guidelines*. In 2005, the GBI licensed Green Globes™ for use in the United States, and this license allows the GBI to promote and further develop Green Globes™ in the United States. With this license, the GBI has adapted Green Globes™ for use in the U.S. commercial building market by referencing U.S. codes and standards, conversion of metric to English units, and incorporating tools such as the U.S. Environmental Protection Agency (EPA) Target Finder.

2.4.3 Green Globes™ Rating System

The Green Globes™ rating system is an interactive Web-based system that includes a self-assessment tool, rating system for certification of the building as a green building, and guide for enhancing a project's sustainability based on the outcome of the self-assessment. The self-assessment tool can be used independently of the rating system if the project owner does not want to pursue third-party verification and certification.

As shown in Figure 2-2, the Green Globes™ rating system is broken down into seven categories and associated number of points achievable by each category as shown in Table: 2-3

Table 2-3

	Green Globes™ Rating Category	Points	Percent
A	Project Management	50	5.0
B	Site	115	11.5
C	Energy	360	36.0
D	Water	100	10.0
E	Resources, Building Materials, & Solid Waste	100	10.0
F	Emissions & Other Impacts	75	7.5
G	Indoor Environment	200	20.0
TOTAL		1000	100.0

		PATH A		PATH B	
A.	**PROJECT MANAGEMENT**				
A.1	Integrated Design20				
A.2	Environmental Purchasing	5			
A.3	Commissioning — Documentation	20			
A.4	Emergency Response Plan	5	**50**		
B.	**SITE**				
B.1	Site Development Area	45			
B.2	Reduce Ecological Impacts	40			
B.3	Enhanced Watershed Features	15			
B.4	Site Ecological Improvement	15	**115**		
C.	**ENERGY**	**PATH A**		**PATH B**	
C.1	Energy Consumption	110		110	
C.2	Energy Demand Minimization	135		135	
C.3	"Right Sized" Energy Efficient Systems	N/A		110	
C.4	Renewable Sources Of Energy	45		45	
C.5	Energy-Efficient Transportation	70	**360**	70	**470**
D.	**WATER**				
D.1	Water	40			
D.2	Water Conservation Features	40			
D.3	Reduce Off-Site Treatment Of Water	20	**100**		
E.	**RESOURCES, BUILDING MATERIALS, & SOLID WASTE**				
E.1	Materials With Low Environmental Impact	40			
E.2	Minimized Consumption & Depletion Of Material Resources	30			
E.3	Re-Use Of Existing Structures	10			
E.4	Building Durability, Adaptability, & Disassembly	10			
E.5	Reduction, Re-Use, & Recycling Of Waste	10	**100**		
F.	**EMISSIONS & OTHER IMPACTS**				
F.1	Air Emissions	15			
F.2	Ozone depletion & Global Warming	30			
F.3	Contamination Of Sewer Or Waterways	12			

Figure 2-2 The Green Globes™ Rating System.

	F.4	Land & Water Pollution	9	
	F.5	Integrated Pest Management	4	
	F.6	Storage For Hazardous Materials	4	**75**
G.	**INDOOR ENVIRONMENT**			
	G.1	Effective Ventilation System	60	
	G.2	Source Control For Indoor Pollutants	45	
	G.3	Lighting Design & Integration Of Lighting Systems	45	
	G.4	Thermal Comfort	25	
	G.5	Acoustic Comfort	25	**200**
TOTAL POINTS				**1000**

Figure 2-2 (*Continued*)

2.4.4 Green Globes™ Self-Assessment

The Green Globes™ rating system can be accessed and updated throughout the design process by the design team to ensure that project information is always up to date and reflects the current state of the design. However, formal self-assessments are completed online at the completion of the schematic design and construction document phases of the building project. At these points in the project, the Green Globes™ rating system provides the project team with feedback regarding how the current project rates in relation to what it could be based on industry standards and best practices. The Green Globes™ rating system also provides the project team with suggestions as to how the project's sustainability could be improved.

2.4.5 Green Globes™ Certification

The GBI provides third-party certification of green buildings for owners who want to pursue certification using the Green Globes™ rating system. Unlike the USGBC's LEED™ rating systems, the GBI's rating system is based on the percentage of points achieved by the project based on the number of applicable points rather than the actual number of points achieved. The GBI's percentage-based rating system is designed to recognize that not all point categories and associated green building strategies are applicable to all building projects. The total number of possible points against which the project will be measured for certification varies based on the nature of the project.

To illustrate the GBI's percentage-based rating system, consider subcategory E.3 in Figure 2-2, which addresses the reuse of existing structures and has 10 points associated with it. If there is no existing structure on the site of the building project being evaluated, the ten points available in this subcategory are not included in the total number of points possible, making the maximum points possible for certification 990 rather than 1,000. In comparison, as can be seen from Figure 2-1, the USGBC's LEED™-NC criteria includes three points in Materials and Resources (MR) Category Credit C1 out of 69 points possible for existing building reuse. If there is no existing structure on the site, there is no adjustment to the points possible.

With the Green Globes™ rating system, the building can earn up to four globes based on the percentage of applicable points earned as shown in Table 2-4. One globe is the lowest Green Globes™ rating, and that requires that the building project achieve a minimum of 35 percent of the applicable points. As can be seen from the following table, the number of globes awarded to a building project increases with the percentage of points earned.

In addition to the percentages associated with each level of Green Globes™ certification, the GBI also provides a qualitative description of the significance of the rating earned by the building as follows:

Demonstrates movement beyond awareness and commitment to sound energy and environmental design practices by demonstrating good progress in reducing environmental impacts.

Demonstrates excellent progress in achieving ecoefficiency results through current best practices in energy and environmental design.

Demonstrates leadership in energy and environmental design practices and a commitment to continuous improvement and industry leadership.

Reserved for select building designs that serve as national or world leaders in energy and environmental performance. The project introduces design practices that can be adopted and implemented by others.

Table 2-4

Certification Level	Percentage of Points Required
1 Globe	35–54
2 Globes	55–69
3 Globes	70–84
4 Globes	85–100

2.4.6 Green Globes™ Verification Process

Green Globes™ certification of a building occurs through a two-stage verification process. The first stage of the verification process occurs when the construction documents are complete. At this stage, the project team completes an online questionnaire, and then the questionnaire responses are verified by the GBI against the documentation generated by the project team during the design process. The design documentation includes not only the construction documents but also the results of analyses and simulations conducted in the course of the design and used to establish benchmarks against which the design is evaluated. Upon successful completion of the first-stage design document review, the project team receives a certificate recognizing its sustainable design.

The second stage in the Green Globes™ verification process occurs after building construction is complete. A GBI representative visits the project and inspects it to verify that the project was completed in accordance with the construction documents. In addition to the construction documents, other documentation such as waste disposal, projected energy use, and life-cycle modeling are also reviewed. Based on the inspection, the GBI determines the Green Globes™ rating that will be awarded, issues a plaque recognizing the project team's accomplishment, and allows the owner to publicize its green building rating.

2.4.7 Information on the Green Globes™ Rating System

Information on Green Globes, the current Green Globes™ rating system, registration and fees for using the Green Globes™ system, as well as Green Globes™ certification can be obtained from the GBI. Contact information is as follows:

The Green Building Initiative
2104 SE Morrison
Portland, Oregon 97214
Telephone: (877) 424-4241
Fax: (503) 961-8991
E-mail: info@thegbi.org
Web site: www.thegbi.org

The GBI provides both online and classroom training on Green Globes™. Information about this training can be obtained on their Web site.

2.5 ADDITIONAL RATING SYSTEMS

There are additional green building rating systems as described in a study performed by the Pacific Northwest National Laboratory in 2006 [Fowler 2006]. The study identified several other applicable rating systems in addition to the USGBC's LEED™ and the GBI's Green Globes discussed in this chapter after screening available rating systems for use in the U.S. construction market based on relevance, measurability, applicability, and availability. These rating systems included BREEAM, which is the forerunner of Green Globes as discussed in this chapter, the Japan Sustainable Building Consortium's (JSBC) Comprehensive Assessment System for Building Environmental Efficiency (CASBEE), and the Green Building Assessment Tool (GBTool™), which was developed as part of the Green Building Challenge by Natural Resources Canada.

2.6 CASE STUDY

McGough Construction

Bloomington Central Station, Minnesota

The "urban village" and the "transit-oriented development" have enjoyed growing popularity in recent years, as Americans have become increasingly fed up with the hassles of commuting, the sterility of the strip-mall environment and the explosive costs of automobile transportation. Urban villages, which are compact, pedestrian-friendly mixed-use neighborhoods, and transit-oriented developments ("TODs")–urban villages accessed by mass-transit–have become the principal models for rejuvenating urban and suburban landscapes. As TODs continue to gain widespread acceptance as the offspring of urban villages, developers look to enhance them by incorporating "eco-friendly" construction–"green" building–into their planning.

The first phase of the Bloomington Central Station development, *Reflections at Bloomington Central Station*, is a 263-unit condominium that features a variety of sustainable construction strategies, including extensive use of recycled building materials, low-emission indoor paints, twice-filtered air, high-efficiency irrigation and stormwater management, and "daylighting"—the use of natural light to reduce the need for electric light while reducing solar heat gain and glare. In part because of its pioneering construction approach, Reflections received the *Minneapolis–St. Paul Business Journal*'s award as "New

Residential Condo (Over 100 Units)," as well as "Best Overall Winner" for 2005. Reflections is one of the first for-sale multifamily residential developments in the nation to receive LEED™ Certification.

Figure 2-3 Photo courtesy of George Heinrich.

Reflections at Bloomington Central Station, situated 330 feet from the Bloomington Central LRT station, qualifies for the LEED™ Public Transportation Credit.

Stormwater management at Reflections condos. Moreover, the site's highly engineered stormwater management system entitles the project to claim multiple credits. The water-efficient landscape design reduces typical irrigation needs by one half. In an effort to reduce the "heat island" effect, there is no surface parking: all of the project's parking is below grade, with a large portion sheltered by an intensive green roof.

The "tight" construction site at Reflections and its impact upon recycling. A compact construction staging area challenged recycling efforts. However, Reflections will receive the Construction Waste Management goal of diverting one-half of all materials, an illustration of the dedication of McGough personnel to accommodating space constraints and learning new waste management procedures.

Reflections condos and the use of recycled materials. Because the building is composed predominantly of concrete and glass, the high fly-ash content in the concrete and recycled aluminum in the curtainwall assembly entitles Reflections to claim its 5 percent Recycled Content goal. Another major contributor is the flooring sound-control underlayment, which is composed of 100 percent postconsumer recycled rubber from car tires. Other recycled materials include rebar, metal framing, insulation, drywall, and steel.

"Daylighting" techniques at Reflections. The efficient sizing of Reflections condominiums—650 to 1,000 square feet—combined with the floor-to-ceiling glass curtainwall, translates into 97 percent of the regularly occupied spaces meeting the LEED™ daylighting requirement, and 98.9 percent of the spaces providing access to views.

Inoperable windows at Reflections produces a marketing benefit for prospects with respiratory issues. Perhaps the most notable quality of Reflections construction is the fact that the windows are not operable—a rarity in condominium design. The high-performance mechanical system is designed around this factor, which ensures that the air inside the structure is purer than outdoor air: the latter is filtered three times before entering the living units, which are continuously exhausted and average one complete air change per hour. Complementing the high level of indoor air quality, 100 percent of the paint in the building meets the LEED™ standard for low-VOC content. In addition, an Indoor Air Quality Management Plan was in effect for the duration of construction.

2.7 REFERENCES

Fowler, K. M., and Rauch, E. M., *Sustainable Building Rating Systems Summary*, Pacific Northwest National Laboratory Operated by Battelle for the U.S. Department of Energy, Contract DE-AC05-76RL061830, July 2006.

Japan Sustainable Building Consortium (JSBC), *Comprehensive Assessment System for Building Environmental Efficiency* (CASBEE), www.ibec.or.jp/CASBEE/english/index.htm, August 22, 2007.

National Resources Canada, *Green Building Assessment Tool* (GBTool™), www.sbc.nrcan.gc.ca/software_and_tools/gbtool_e.asp, August 21, 2007.

chapter 3

Understanding Green Project Requirements

3.1 INTRODUCTION

The key to reducing risk associated with green construction is to thoroughly understand the project requirements. This understanding starts with the owner's bid documents, which will be the basis for the contractor's bid and includes the owner's bid requirements, contractual requirements, and technical factors that will impact the contractor's scope of work and risk. This chapter builds on previous chapters by pointing out where in the construction documents the contractor can expect to find green construction requirements and their possible impact on cost, schedule, and productivity throughout the construction process from mobilization through commissioning. Green construction requirements that address technical and administrative requirements are cross-referenced with both the 1995 and 2004 editions of the Construction Specifications Institute's (CSI) *MasterFormat*™ that are currently in use.

3.2 IMPACT OF GREEN REQUIREMENTS

As discussed in this chapter and throughout the remainder of this book, green project requirements can impact all aspects of the construction process as well as the contractor's costs, schedule, and productivity. There is often a misconception that green building construction impacts only the design and it is business as usual for the contractor. This is not the case. First, the green design must be implemented by the contractor in order for the green building to become a reality. Green design requirements will impact the contractor's procurement process as well as the construction and project closeout. In addition, many requirements on green building projects will impact how the contractor and its subcontractors carry out the work at the project site. These requirements include site layout and use, construction waste management, material storage and protection, indoor air quality during construction, and others. Therefore, the contractor must understand the green project requirements, which include not only the contract document requirements but also the requirements of any third-party green building rating system that is used on the project (as discussed in Chapter 2).

3.3 GREEN PROJECT DELIVERY

3.3.1 Importance of Green Project Delivery

A green construction project can be organized in a variety of ways to take it from concept to completion for the owner. How a construction project is organized is often referred to in the construction industry as a project delivery system. The owner typically decides how a green construction project will be organized in the early planning stages of the project. The project delivery system is so important to the contractor on a green building project because the project organization determines the contractor's involvement in the project, which in turn affects its scope of work and risk. This section discusses some common project delivery systems that are used on green construction projects. There is no one right way to organize a green construction project. Each project is unique, and the project delivery system should be tailored to the specific construction project as well as the owner's capabilities.

The information on project delivery systems contained in this chapter is only intended to provide an overview of some of the more common project delivery systems and how they relate to green construction. The discussion is intended to give the reader a framework for discussions that follow in later chapters about specific aspects of green construction and the green construction process that are impacted by the project delivery system. For instance, Chapter 5 addresses managing green design on design-build projects. For a more complete discussion of project delivery systems, the reader is directed to *Project Delivery Systems for Construction*, published by The Associated General Contractors of America.

3.3.2 What Is a Project Delivery System?

A project delivery system defines how a construction project will be organized in order to take it from the owner's concept to physical reality. This organization needs to be matched to the owner's in-house design and construction capabilities as well as the unique characteristics of the project. The way in which the project is organized will affect how efficiently and effectively it can be designed and built.

Unlike most business organizations, organizations formed to design and build construction projects are temporary. In most cases, the owner, designers, and constructors come together to complete a construction project, and then after it is complete, they disband and go their separate ways. In fact, most of the people and organizations that are involved in a construction project are not involved throughout the entire design and construction process. Specialty consultants perform their portion of the design and then move on to other projects, and specialty contractors move on and off the construction site as required by the construction sequence. In addition, people within the organizations change throughout the project because different expertise is needed in the various project phases, people get involved in other projects, or they leave the organization. This is a challenge on any construction project, but it is especially challenging on green construction projects where success depends on continuity and a common focus on achieving the owner's sustainability goals for the construction project throughout design and construction.

Project delivery systems are very important in construction because they provide a common framework that people and organizations associated with the construction project understand and can work within. A project delivery system assigns authority and responsibilities to people and organizations as well as defines the relationship between them. These responsibilities, authorities, and relationships are typically defined in the contract documents that provide the blueprint for the project delivery system. Again, the project delivery system selected by the owner for a green project is very important for the contractor because it will impact its role, responsibilities, and risk on the project.

3.3.3 Project Delivery System Categories

Several project delivery systems are used on construction projects today. These project delivery systems can be grouped into the following two categories:

- Construction Manager/General Contractor
- Design-Build

The following sections will briefly describe each of these project delivery system categories and the more common project delivery systems that fall into each category. A more detailed discussion of each of the individual project delivery systems will follow.

Construction Manager/General Contractor. The construction manager/general contractor category includes several project delivery systems that are commonly used on green construction projects. Project delivery systems that fall into this category require that the owner contract directly and separately with both the designer and the general contractor to complete the project. The two most common project delivery systems that fall into this category are the following:

- Design-Bid-Build
- Construction Manager At-Risk

The major difference between the design-bid-build and construction manager at-risk is when and how the contractor is brought into the project delivery process. With design-bid-build project delivery, the design team completes the design first, and then the owner contracts with the general contractor to build the project based on that design. With design-bid-build, the general contractor is usually selected based on low price through a competitive bidding process.

When using the construction manager at-risk project delivery system, the contractor is often referred to as the construction manager. The construction manager is usually selected by the owner based on qualifications or best value that address both qualifications and price. Typically, the contractor is involved in the green construction project during project planning and design. In addition to being responsible for constructing the green project, the contractor also assists the owner and its design team throughout the planning and design process. During the planning and design process, the contractor can provide valuable input and assistance regarding cost, schedule, and construction issues.

Design-Build. The design-build project delivery system requires the owner to contract with only one entity for both the design and construction of the project: the design-builder. Typically, the design-builder will be a contractor on a design-build green construction project. However, the design-builder could also be an architect, developer, or other entity. Like a construction manager, the design-builder can be selected by the owner based on qualifications, price, or best value that addresses both qualifications and price.

3.4 GREEN DESIGN-BID-BUILD

Figure 3-1 illustrates the design-bid-build project delivery system. Under this project delivery system, the owner contracts separately with both an architect and a general contractor to get the project completed. The owner first contracts with the architect to perform the design. Once the design is complete, the owner typically bids out the project to either a select or open list of qualified general contractors. As notedpreviously, green building projects are as much about construction as they are about design. The disadvantage of design-bid-build on green building projects is that the contractor is not involved in the project until after the design is complete. Making changes to the design that will reduce construction costs and waste as well as increase building efficiency after occupancy is difficult at this point. In addition, the owner loses the benefit of the contractor's expertise in performing value analyses and constructability reviews during the design process as well as cost estimating and scheduling.

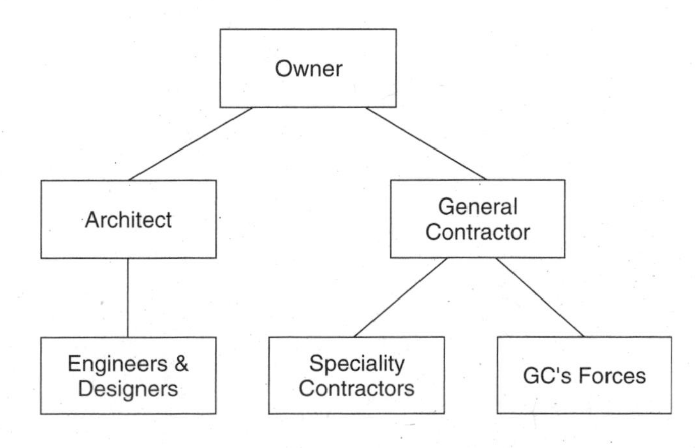

Figure 3-1
Design-Bid-Build Project Delivery System.

3.5 GREEN CONSTRUCTION MANAGER AT-RISK

Construction management is a broad term covering a variety of project delivery methods that all include a construction manager as part of the project team to oversee scheduling, cost control, constructability, project management, building technology, bidding or negotiating construction contracts, and construction. While the term implies the management of construction only, construction managers may also assist the owner during the planning and design phase of the green construction process. Construction management is very appropriate for green construction projects because green building projects are typically more complex, require closer monitoring of budget and schedule throughout the design and construction processes, can benefit from the input of the construction manager during design, usually require extensive coordination of consultants and specialty contractors, and often have more extensive commissioning and closeout requirements.

The construction manager at-risk project delivery system is illustrated in Figure 3-2. The construction manager is brought in at the beginning of the project as the owner's advisor and, like the general contractor in Figure 3-1, the construction manager contracts with and coordinates the specialty contractors. The construction manager can contract with the owner to be paid for its services based on a fixed fee, percentage of construction cost, or percentage of construction cost with a guaranteed maximum price.

With design-bid-build, the general contractor usually does not get involved with the project until after the design is complete, and the owner either bids out the project or contacts the general contractor to negotiate the work. As a result, the owner is forced to rely on the architect for prebid cost estimates and scheduling. The architect's expertise is typically not in material

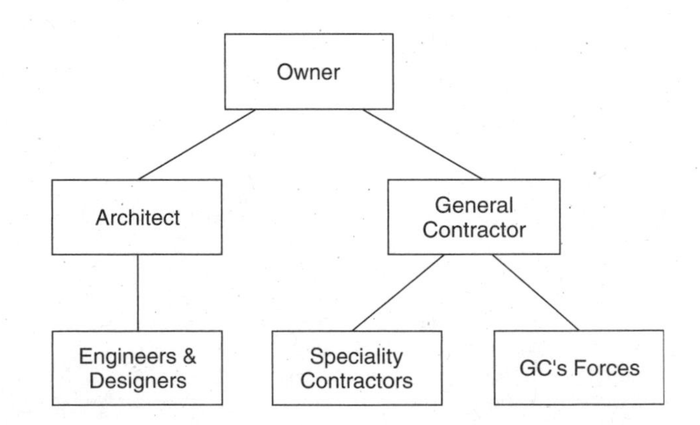

Figure 3-2 Construction Manager – At-Risk Project Delivery System.

and equipment procurement or construction means and methods. Input to the design process by a knowledgeable construction manager can be very valuable to the owner during the design process on a green building project. Constructability reviews and value analyses, coupled with ongoing budget and schedule reviews by a construction manager, could result in better material and equipment selection, reduced waste, increased construction efficiency, and lower life-cycle costs for the project.

3.6 GREEN DESIGN-BUILD

3.6.1 Design-Build Overview

With design-build, only one organization is responsible for both the project design and construction. Under design-build, the owner has only one contract for the complete delivery of the project that includes both design and construction. The design-build project delivery system is illustrated in Figure 3-3.

From Figure 3-3, it can be seen that the owner needs to administer only one contract for the construction, renovation, or expansion project. In addition, one entity is responsible for both design and construction, and the owner no longer finds itself between the architect and contractor. This means that if design problems impact construction, the design-builder must resolve them within the design-build team, and the owner does not have to get involved. Furthermore, if the performance of the building systems does not meet the

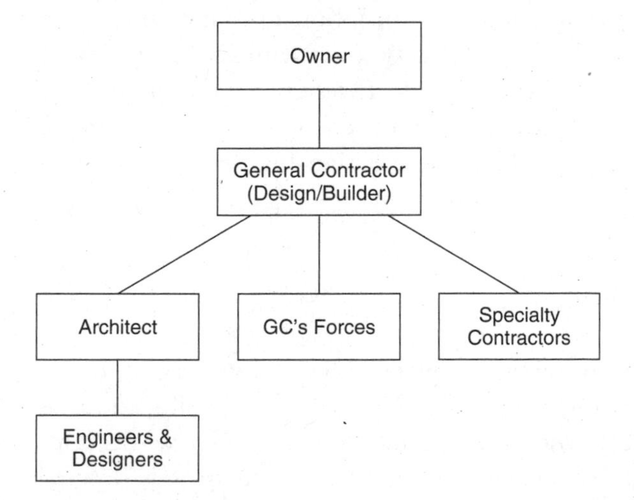

Figure 3-3 Design-Build (GC as Design-Builder) Project Delivery System.

measurable performance criteria agreed upon at the beginning of the project, then it is not a question of whether it is a design or construction issue that the owner needs to resolve. Instead, the design-builder again needs to resolve the issue within the design-build team, and the owner does not need to be involved. The only time that the owner should be faced with a change order on a design-build project is if it makes a change to the agreed-upon project criteria or if the design-builder encounters conditions beyond its control that it could not have reasonably anticipated at the beginning of the project.

As a result of the designers and constructors all working together on the same team on a design-build project, the owner also gets the advantage of construction expertise throughout the design process and design expertise throughout the construction process. Similar to construction management at-risk, having designers and constructors working together as a team can result in shorter construction schedules and increased construction value for the owner. However, the even closer collaboration between the contractor and design team on a design-build project can be even more valuable on a green building project. In addition, with design-build, the owner has only one contract to manage, which reduces the owner's administrative burden and makes project management much simpler from the owner's viewpoint. The design-builder would have sole point responsibility for achieving the owner's green project goals and green building certification.

3.6.2 One-Stage Design-Build Process

Under the one-stage design-build process, the owner selects a design-builder to take the green construction project from inception to completion. The design-builder can be selected based on low price, qualifications, or a combination of both. Often the selection of a design-builder involves a request for qualifications (RFQ), request for proposals (RFP), shortlisting, interviews, and evaluations. Once selected, the design-builder begins design and construction of the project.

3.6.3 Two-Stage Design-Build Process

Public design-build projects are often completed using a two-stage process. The first stage involves selecting a design-builder based on qualifications as well as estimated total project cost. The selected design-builder works with the owner to define its needs and expectations if it has not adequately defined them in its program, to establish measurable performance objectives for the completed project, and to take the design to a predetermined point. At this point, the design-builder has the opportunity to reconfirm its original estimated project cost and schedule for the project, and the owner has the opportunity to evaluate the design to date.

If both the design-builder and owner find that they are in agreement at the end of this first stage, they can negotiate and execute a contract for the second stage that involves completing the design and construction of the project. If the design-builder and owner do not want to proceed together with the completion of the project, the owner pays the design-builder for work performed to date and then can use the design documents developed to date to negotiate with another design-builder or bid out the design with multiple design-builders for completion.

3.6.4 Design-Build Variations

Several design-build variations can be used for a green construction project. With the classic design-build process described previously, the design-builder takes the owner's criteria, which may include the requirement that the building be certified as a green building by a third-party organization, and designs and constructs the building to meet those criteria. Alternately, the owner and its architect can take the design to some point, such as the completion of schematic design where building operational requirements, major building systems, and building materials have been defined. At this point, the design-builder is selected to complete the design and build the building. Terms used to describe this variation on the classic design-build project delivery system are design-assist or draw-build. Selection is usually based on low price using the competitive bid process, and the successful design-builder completes the design and constructs the green project.

These variations provide some unique challenges for the contractor acting as a design-builder on a green building project. For one, the contractor is not involved in the planning and early design stages of the green building project, so the contractor's ability to provide valuable input to the owner and design team based on its expertise and experience is limited as discussed previously with design-bid-build project delivery. Another consideration is the risk that the contractor is assuming on a green building design-build project using draw-build or design-assist. The major building systems and equipment are typically set during the schematic design phase of the project, and under these design-build variations, the contractor is only responsible for detailing the systems. If the systems do not function as planned during the startup and commissioning process (as discussed in Chapter 4), the question becomes who is responsible for the system performance—the designer who performed the system design or the contractor who detailed and installed the system? Sorting this out, especially when a third-party green building rating system is involved, can be very difficult and time consuming.

3.6.5 Managing the Green Design Process

The contractor who is assuming the role of design-builder on a green project needs to understand how to manage not only construction but also the design process. Green building design is more complex than what is usually required for a traditional building. Sustainable design typically requires the evaluation of alternative materials and systems by the designer as well as building modeling and simulation to ensure that the constructed building will meet the owner's performance requirements and conform to industry standards, energy codes, and green rating systems. Chapter 5 addresses managing green design for design-build projects and should also be useful to the construction manager in working with the owner and designer during the planning and design process for a green project.

3.7 CONTRACTOR SELECTION METHODS

Common methods for selecting a contractor were mentioned for each of the project delivery systems discussed previously. The selection process is important and can have a major impact on the success of a green construction project. The three basic methods that an owner can use to select a contractor for a green building project are as follows:

- Price-based selection
- Qualifications-based selection
- Best-value selection

3.7.1 Price-Based Selection

The contractor can be selected based solely on low price for the project scope of work. Price-based selection is typically accomplished through a bidding process, although selection based on low price can also be done through negotiation. Pure price-based selection is typically used on design-bid-build to select a general contractor based on low bid. Price-based selection can also be used to select a design-builder for a design-build project. However, selecting a design-builder based on low bid can be very risky for the owner, because the project is not well defined at the point when a design-builder is selected, and the owner may not get the project it wants. The most common method for selecting a general contractor for a public design-bid-build green construction project is price-based selection using a competitive bidding process.

3.7.2 Qualifications-Based Selection

With qualifications-based selection, the contractor is selected solely on the qualifications to perform the work. Qualifications-based selection can consider factors such as past experience on green building projects as well as the expertise and capabilities of the project team. Qualifications-based selection usually involves a negotiated price for the construction of the project and may not be able to be used on a public green construction project. Qualifications-based selection is most often used in the selection of a contractor for a project using construction manager at-risk, design-build, or construction manager-agency. However, qualifications-based selection can also be used to select a general contractor after the project design is completed instead of bidding the project and selecting the general contractor with the lowest bid as is typically done in design-bid-build.

3.7.3 Best-Value Selection

Best-value selection is any process where the successful contractor is selected based on both price and qualifications. Using best value. the owner weighs each competing firm's price and qualifications and makes a selection. For a green design-build project, the conceptual design would also be a major factor in the selection process. Best-value selection is most often used to select a contractor for a construction manager at-risk or design-build, which requires services beyond constructing the project in accordance with construction documents. Best value can be used on public construction projects as long as the selection process meets the requirements of the state and other approving authorities.

3.8 GREEN SCOPE OF WORK IMPACT

The prebid analysis of a project is an extremely important part of the bid preparation process. Without a detailed and comprehensive analysis of the bid documents, the contractor cannot possibly understand its project scope and develop a reliable plan for performing the work that will serve as the basis for the bid estimate. The planned means, methods, and scheduling of the project activities can have a significant impact on estimated project costs. Project green requirements can impact the design, procurement, construction, commissioning and closeout, as well as postoccupancy warranty and operational verification.

3.9 IDENTIFYING GREEN REQUIREMENTS

The owner's green project requirements can be included in a variety of places in the project contract documents. This is why it is imperative that the contractor conducts a thorough and detailed review of the owner's request for bid and contract documents before bidding, looking specifically for green requirements for which the contractor will be responsible. In general, green requirements will be found in the project specifications, as will be discussed in the remainder of this chapter. However, they may also be incorporated into the project requirements and the owner-contractor agreement by reference to local laws, codes, or standards that include green building requirements or require that the building be certified or certifiable as a green building.

3.10 SPECIFICATIONS

3.10.1 Specifications Defined

The Associated General Contractors of America (AGC) Document No. 200 does not define the term *specifications*, but the American Institute of Architects (AIA) Document A201™ defines *specifications* as follows:

> The Specifications are that portion of the Contract Documents consisting of the written requirements for materials, equipment, systems, standards and workmanship for the Work, and performance of related services.

Therefore, the project specifications are key to understanding green construction requirements and must be thoroughly reviewed, analyzed, and understood before bidding.

3.10.2 Specification Types

There are three types of specifications as follows:

- Descriptive specifications
- Prescriptive specifications
- Performance specifications

Descriptive Specifications. Descriptive specifications are the most common way that materials and equipment are detailed in construction, including green building projects. A descriptive specification describes the material or equipment in terms of its materials, physical construction, and performance characteristics. Descriptive specifications typically use industry standards and test criteria as a shorthand method of establishing minimum requirements for the material or equipment. Descriptive specifications also often identify one or more manufacturers as acceptable producers of the material or equipment, further defining the level of quality desired and usually including an "or equal" clause, opening it up further to other manufacturers that are not listed but that can provide products of equivalent quality.

Prescriptive Specifications. Prescriptive specifications require the contractor to supply a specific material or equipment that is produced by a specific manufacturer. Prescriptive specifications are usually used where there is a need to match a specific manufacturer's equipment, such as in the case of adding cubicles to an existing medium-voltage metal-clad switchgear lineup. Another instance when prescriptive specifications are used is when the contractor is adding to, modifying, or extending an existing system, such as a proprietary fire alarm or security system where only equipment designed, tested, and listed for use on that proprietary system may be used. Prescriptive specifications typically lock the contractor into a single-supplier arrangement that can result in increased cost and possible schedule impacts. Prescriptive specifications are also used on green building projects for the same reasons they are used on traditional building projects.

Performance Specifications. Both descriptive and prescriptive specifications define exactly what materials and equipment are to be procured by the contractor and how these materials and equipment are to be installed. With descriptive and prescriptive specifications, the owner takes responsibility for the performance of the installation as long as the contractor installs the materials and equipment in accordance with the contract documents.

Performance specifications, however, only specify how the system is to perform after installation. Restrictions regarding materials and equipment used as well as the means and methods of installation are kept to a minimum by the owner. Meeting the specified performance requirements is the contractor's responsibility. Within the confines of the contract documents, the contractor is free to select its own materials and equipment and installation method to achieve the owner's stated performance requirements. With a performance specification, the contractor takes responsibility for the performance of the installed system and guarantees that it will meet the owner's stated performance criteria.

Performance specifications are used on design-build projects as well as on traditional design-bid-build and negotiated projects where proprietary systems are required and components are not interchangeable. Examples of systems that are typically procured through performance specifications are fire alarm systems, security systems, nurse call systems, and others. In these cases, it is easy to specify the desired performance of the installed system, but it would be impossible to specify components and installation using a descriptive or prescriptive specification because each system is different. If the specification were written around one manufacturer's system, it would most likely exclude all others, making the intended performance specification a prescriptive specification by default.

In addition to specifying materials, equipment, and systems using performance specifications on green building projects, achieving certification or being certifiable as a green building via a third-party rating system can also be a performance specification. In other words, the contract documents could require that the contractor meet the requirements for green building certification, but no direction is provided as to how the contractor is to achieve this requirement. This is most often encountered on design-build green projects where the contract documents may give the contractor full discretion as to how it attains the green building certification. Or there may be a set of owner-specified requirements that the contractor must meet, and the remaining requirements needed to meet the green building requirement are left to the contractor. This could also occur on a design-bid-build project, where the contract documents explicitly require the contractor to achieve a green building certification or implicitly require the contractor to meet applicable codes and laws, which may require the building to be certified or certifiable as a green building.

3.10.3 Conflicting Green Specifications

In reviewing the specifications on a green building project, the contractor needs to look for conflicting green specifications. Green building rating systems, green product criteria, certifications, and labeling, as well as codes and standards, are constantly changing. The design team may require that a product meet a specified green material criteria in the specification and then list specific products that may or may not meet the green product requirements. Although the burden should be on the design team to ensure that the specified product meets the required green criteria, it is best that the contractor verify that the product meets the required green criteria before purchasing and, if possible, before bidding. Even though it is the design team's responsibility to ensure that named products in the specification meet the green criteria, the time spent by the contractor verifying that the product purchased and installed does meet these criteria may save a lot of time, effort, and cost in resolving the issue, performing the necessary rework, and negotiating a change order.

3.10.4 Mixed Green Specifications

The contractor should be careful to avoid mixed green specifications. A mixed specification contains the elements of both a descriptive or prescriptive specification and a performance specification. Mixed specifications are usually encountered primarily because the person drafting the specification is inexperienced and does not understand its implications. In the case of a mixed specification, the contractor is told what materials and equipment must be used and how those materials and equipment must be installed, just like in a descriptive or prescriptive specification. In addition, the contractor is also responsible for the system meeting owner-specified performance criteria as in a performance specification. An example of a mixed specification might be the material specification for a particular carpet chosen for its aesthetics, which conflicts with performance criteria that the interior finishes meet the indoor air-quality requirements of a third-party green building rating system.

This type of specification can result in a no-win situation for the contractor if the material, equipment, or system does not perform as specified, although it was installed in accordance with the owner's requirements. Mixed specifications should always be avoided because they are fertile ground for disputes and increase the contractor's risk. The fact that courts may only hold the contractor to the prescriptive portion of a mixed specification is of little consolation when the cost in time, money, and loss of owner goodwill involved in resolving a dispute like this is considered.

3.11 CSI MASTERFORMAT™

3.11.1 CSI MasterFormat™ Background

The Construction Specifications Institute (CSI) is a professional organization whose purpose is to promote better organization and communication of construction project information. The MasterFormat™ is a list of numbers and titles for organizing information about construction requirements, products, and activities into a standard sequence. The current edition of the CSI MasterFormat™ is the 2004 edition, but because of the extensive changes that were incorporated into the 2004 edition, the 1995 edition is still frequently used. The contractor should be aware of where project green requirements are typically specified in both editions.

3.11.2 2004 MasterFormat™ Groups

The 2004 CSI MasterFormat™ divisions are divided into the following two groups:

- Procurement and Contracting Requirements Group
- Specifications Group

Procurement and Contracting Group – Division 00. The Procurement and Contracting Group consists of only Division 00 entitled Project Procurement and Contracting Requirements. Division 00 is not part of the technical specifications for a construction project. The purpose of Division 00 is to provide a consistent framework for providing bidding and contracting information about the project. This division provides contractors with information about the project, bidding requirements, and contracting requirements. Even though Division 00 provides no technical information about the project, the contractor must be familiar with the information in this division. Specific information provided in this division that the contractor needs to be aware of includes the project scope, scheduled prebid meetings, bidding requirements, information available to bidders during the bidding stage, required bid information and forms, and the owner-contractor agreement that will be used to contract for the work.

For example, the 2004 CSI MasterFormat™ contains the following two Level 3 sections under Certificates and Other Forms, which is Level 2 Section 00 62 00 that explicitly address green project requirements:

- 00 62 23 Construction Waste Diversion Form
- 00 62 34 Recycled Content of Materials Form

Specifications Group. The Specifications Group contains all of the remaining 33 divisions that comprise the 2004 CSI MasterFormat™. These divisions are divided into five subgroups that address specific types of construction materials and equipment, as well as specific systems and construction market categories. The five subgroups that comprise the Specifications Group are as follows:

- General Requirements Subgroup (Division 01)
- Facility Construction Subgroup (Divisions 02–14)
- Facility Services Subgroup (Divisions 20–29)
- Site and Infrastructure Subgroup (Divisions 30–39)
- Process Equipment Subgroup (Divisions 40–49)

The focus of this book is building construction, so the Site and Infrastructure Subgroup and the Process Equipment Subgroup will not be addressed.

3.11.3 General Requirements Subgroup – Division 01

The General Requirements subgroup consists solely of Division 01 entitled Project General Requirements. Division 01 provides the general requirements for the project and applies to all other specification divisions that follow.

Among other things, Division 01 provides the general requirements for the following:

- Contract Administration
- Contract Performance
- Material and Equipment
- Quality Assurance and Control
- Payment Procedures
- System Startup and Testing
- Contract Closeout

The general requirements contained in Division 01 can have a significant impact on the contractor's work. The contractor needs to thoroughly understand the Division 01 requirements and realize that even though these requirements may not be repeated in other technical divisions, the requirements are part of the project and must be addressed. Green requirements may be included in this division that impact other divisions that the contractor plans to subcontract. Therefore, in developing the scope documents for subcontractors, the contractor needs to make sure that Division 01 requirements are included in the subcontractor's scope and that they are aware of the requirements. Dividing the green project scope and preparing bid documents for subcontractors are covered in detail in Chapter 6.

There are several explicit references in Division 01 of the 2004 CSI MasterFormat™ to green project requirements. These specific Level 3 sections in Division 01 are as follows:

- 01 33 29 Sustainable Design Reporting
- 01 74 19 Construction Waste Management and Disposal
- 01 78 53 Sustainable Design Closeout Documentation
- 01 81 13 Sustainable Design Requirements
- 01 81 16 Facility Environmental Requirements
- 01 81 19 Indoor Air-Quality Requirements

In addition to explicit references to sustainability requirements in Division 01, several other sections could contain sustainability requirements for a green building.

Section 01 56 00 addresses temporary barriers and contains the following Level 3 sections that could contain requirements for temporary barriers for indoor air quality during construction:

- 01 56 13 Temporary Air Barriers
- 01 56 16 Temporary Dust Barriers
- 01 56 19 Temporary Noise Barriers

Temporary controls are specified in Section 01 57 00 that includes the following subsections:

- 01 57 13 Temporary Erosion and Sediment Control
- 01 57 19 Temporary Environmental Controls
- 01 57 23 Temporary Storm Water Pollution Control

Performance requirements are specified in Section 01 80 00 and include the following subsections:

- 01 81 00 Facility Performance Requirements
- 01 82 00 Facility Substructure Performance Requirements
- 01 83 00 Facility Shell Performance Requirements
- 01 84 00 Interiors Performance Requirements
- 01 85 00 Conveying Equipment Performance Requirements
- 01 86 00 Facility Services Performance Requirements
- 01 87 00 Equipment and Furnishings Performance Requirements
- 01 88 00 Other Facility Construction Performance Requirements
- 01 89 00 Site Construction Performance Requirements

General building commissioning requirements are specified in Level 2 section 01 91 00, which includes the following Level 3 sections:

- 01 91 13 General Commissioning Requirements
- 01 91 16 Facility Substructure Commissioning
- 01 91 19 Facility Shell Commissioning
- 01 91 23 Interiors Commissioning

These general commissioning requirements may be supplemented by specific commissioning requirements in other specification divisions in the Facility Construction Subgroup and Facility Services Subgroup. For example, heating, ventilating, and air-conditioning (HVAC) systems and equipment are specified in Division 23. Level 3 section 23 05 93 specifies required testing, adjusting, and balancing for HVAC systems, and Level 2 section 23 08 00 provides the specific requirements for HVAC commissioning.

3.11.4 Facility Construction Subgroup

The Facility Construction Subgroup includes 13 divisions that address various materials and equipment commonly encountered in residential, commercial, and institutional construction. The Facility Construction Subgroup divisions are numbered consecutively from Division 02 through Division 14. Figure 3-4 provides a list of the Facility Construction Subgroup divisions and compares the 2004 edition of the MasterFormat™ and the previous 1995 edition for these divisions. As can be seen from Figure 3-4, outside of minor

2004 MASTERFORMAT™		1995 MASTERFORMAT™	
NUMBER	TITLE	NUMBER	TITLE
02	Existing Conditions	2	Site Construction
03	Concrete	3	Concrete
04	Masonry	4	Masonry
05	Metals	5	Metals
06	Wood, Plastics, & Composites	6	Wood & Plastics
07	Thermal & Moisture Protection	7	Thermal & Moisture Protection
08	Openings	8	Doors & Windows
09	Finishes	9	Finishes
10	Specialties	10	Specialties
11	Equipment	11	Equipment
12	Furnishings	12	Furnishings
13	Special Construction	13	Special Construction
14	Conveying Equipment	14	Conveying Systems
15	Reserved For Future Expansion	15	Mechanical
16	Reserved For Future Expansion	16	Electrical
17	Reserved For Future Expansion		
18	Reserved For Future Expansion		
19	Reserved For Future Expansion		

Figure 3-4 Facility Construction Subgroup Divisions.

title changes, the materials and equipment addressed in Divisions 02 through 14 remained substantially unchanged.

3.11.5 Facility Services Subgroup

The Facility Services Subgroup is new to the 2004 CSI MasterFormat™ and addresses the mechanical, electrical, and plumbing systems that were previously covered in the 1995 edition by Division 15/Mechanical, Division 16/Electrical, and part of Division 13/Special Construction. The Facility Services subgroup now comprises the following seven divisions as shown in Table 3-1 below:

Table 3-1

Number	Title
21	Fire Suppression
22	Plumbing
23	Heating, Ventilating, Air-Conditioning
25	Integrated Automation
26	Electrical
27	Communications
28	Electronic Safety and Security

Source: Construction Specifications Institute 1996 & 2004.

3.11.6 1995 CSI MasterFormat™ References

In the 1995 edition of the CSI MasterFormat™, green requirements are not explicitly addressed. General requirements that apply throughout the specification by reference will probably be found in Division 1. The architects, engineers, and other designers can then include specific requirements elsewhere throughout the remaining 14 divisions.

Division 1 provides the general requirements for the entire project, including all of the specification sections that follow. Division 1 usually includes a general listing of codes and standards that need to be met on the project as well as specific project requirements that apply universally to all of the technical specification divisions. The contractor needs to review Division 1 in detail to make sure that it understands any general green requirements contained in Division 1 that it will be held to, even though the requirement may not be referenced elsewhere in the specifications.

Green requirements may also be included in Divisions 02 through 16 either directly or by reference, even though the 1995 CSI MasterFormat™ does not have a specific section or subsection for green requirements. When reviewing the technical divisions, the contractor should check all references to other specification sections to ensure that it knows about any green requirements to which the project will be subject.

3.11.7 A Word of Caution

This section has identified some specific CSI MasterFormat™ divisions, sections, and subsections where the green requirements for a project's installation might be addressed in both the current 2004 edition and the previous 1995 edition. However, the CSI MasterFormat™ is only an advisory document, and there is no guarantee that the owner or designer will adhere to either edition's numbering system. Therefore, the contractor needs to make sure that green requirements do not appear somewhere else in the project specifications.

3.12 GREEN REQUIREMENTS ON DRAWINGS

Drawings provide a graphic representation of the scope of work to be performed. AGC Document No. 200 does not define the term *drawings*, but AIA Document A201™ defines *drawings* as follows:

> The Drawings are the graphic and pictorial portions of the Contract Documents showing the design, location and dimensions of the Work, generally including plans, elevations, sections, details, schedules, and diagrams.

Drawings show how the finished work is to appear, how materials and components are to be integrated together, and the dimensions and layout of the work.

Except on small projects where the specifications are included directly on the drawings, the green requirements will usually not be included on the drawings. An exception to this might be a reference in the general drawing notes to the green requirements contained in the specifications.

3.13 CASE STUDY

Austin Commercial

Texas Instruments R-Fab, Texas

RFab is a Texas Instruments (TI) 300 mm silicon wafer manufacturing plant occupying 1.1 million square feet on 92 acres in Richardson, Texas. Built in a brisk 18 months, it is a model for conceiving and planning a sustainable facility from an industry that usually does not have the time to do so.

Microchip plants are energy intensive, using as much electricity as 10,000 homes. Yet in the boom-and-bust world of fab construction there never seems to be time to tinker with energy efficiency: you must build if you're booming; there's no money to design if you're not.

Figure 3-5 Photo courtesy of Texas Instruments.

However, when TI management set a goal of reducing RFab's capital cost by 30 percent, the facilities team set a like goal to reduce operating costs. This created an opening to look at sustainability. A three-day charette with

The Rocky Mountain Institute launched a fab design overhaul that achieved significant energy and water savings without increasing capital cost. They discovered that the same intense process could be used to reduce both capital and operating costs.

Figure 3-6 Photo courtesy of Texas Instruments.

The owner has vision. Because TI knew what they wanted from their sustainability effort they expect to get LEED™ silver certification for at most a 1 percent premium on construction cost, and possibly for no premium.

TI gets into their toolmakers' business. TI asked toolmakers for smaller, cooler and more efficient tools. Beyond improving the tools' performance, savings cascaded into the building: less space required, less heating and cooling, and less process piping. TI pressed whole system design limits and showed how LEED™ can help create a paradigm for wider sustainability efforts.

Hot, thick, clean room smocks get redesigned and facemasks get tossed. Again transcending traditional design boundaries, TI found they could use lighter-weight smocks and eliminate facemasks without harming clean room quality, allowing workers to be more comfortable and reducing cooling requirements (touching on two tenets of LEED™: building occupant comfort and energy efficiency).

Austin Commercial gives its subcontractors a hand. This was an early LEED™ project for Austin and many subcontractors. To get it right, Austin listed qualified materials in bid packages, then audited to ensure the materials were used, stopping to correct where needed so that valuable recycled content, rapidly renewable material, and certified wood points would not be lost after project completion. Austin also advised subcontractors of alternates for regionally sourced materials to keep this point intact. Substituting process and enthusiasm for experience helped ensure that Austin added value.

Vacuum pump efficiency nearly doubled, cascading into reduced chiller and power requirements. The net result was a 20 percent reduction in energy and 35 percent reduction in water use, which saves $4 million annually when running at full capacity.

Dumpster diving results in a 90 percent recycling rate. Austin Commercial stationed a "recycle controller" in the recycling area to ensure that the right materials went into the right dumpsters, occasionally going into the dumpsters to retrieve misplaced material. This kept loads from being disqualified, which would lower the recycling rate and trigger load fees. Recycling saved $50,000, even after paying for the dumpster divers.

Lessons learned. Successful LEED™ projects start with the vision and leadership of the owner. Staffing with enthusiastic, creative project participants helps too.

3.14 REFERENCES

The American Institute of Architects, *Standard Form of Agreement Between Owner and Contractor (Where the Basis of Payment Is a Stipulated Sum)*, AIA Document A101™-1997, 1997.

The American Institute of Architects, *General Conditions of the Contract for Construction*, AIA Document A201™-1997, 1997.

The Associated General Contractors of America, *Project Delivery Systems for Construction*, Second Edition, 2004.

The Associated General Contractors of America, *Standard Form of Agreement and General Conditions Between Owner and Contractor (Where the Contract Price Is a Lump Sum)*, AGC Document No. 200, 2000. [Refer also to pending ConsensusDOCS: General Contracting Number 200 entitled *Owner/Contractor Agreement & General Conditions (Lump Sum)*, 2007.]

The Construction Specifications Institute, *MasterFormat™: Master List of Numbers and Titles for the Construction Industry*, 1995 Edition, 1996.

The Construction Specifications Institute, *MasterFormat™: Master List of Numbers and Titles for the Construction Industry*, 2004 Edition, 2004.

chapter 4

Contracting for Green Construction

4.1 INTRODUCTION

More and more facility owners want their organizations recognized as being environmentally conscious. An important way that these owners are demonstrating their commitment to the environment is through green construction. Green building rating systems (discussed in Chapter 2) allow owners to be openly recognized for their commitment to the environment. These rating systems provide an objective, consistent, and measurable method for determining the degree to which a new or renovated building is a green building. In general, contracting for a green construction project should be no different than any other project. However, the owner's desire to achieve a certain level of building performance or certification for the building may impact the contractor's project scope of work. It is very important that the contractor understand its contractual responsibilities on a green project in order to control its risk and ensure a successful project. This chapter discusses contracting for green construction.

4.2 CONSULT YOUR ADVISORS

The purpose of this chapter is to discuss contracting for green construction. The information provided in this chapter is general in nature and not intended to provide legal, insurance, or bonding advice. This chapter is only intended to provide the contractor with information to assist in negotiations with the owner and discussions with the contractor's attorneys, insurance carriers, and sureties. The contractor should consult its attorneys, insurance carriers, and sureties in any contract, insurance, or bonding matters. It should be noted, however, that green building construction is new to the construction industry, and advisors may not be knowledgeable about the risks associated with green building construction. They should take the time to educate themselves about green building construction and third-party green building rating systems. Until green construction becomes mainstream in the construction industry, risk management should be a team effort.

4.3 CONTRACT PURPOSE

The purpose of any contract is to establish a private set of rules that governs the business relationship between the two contracting parties. When contracting for any construction project, the contractor and the project owner are agreeing to a set of rules that will govern their behavior during construction and through the warranty period that follows. In many ways, a contract can be thought of as a private legal system. The contract not only defines the relationship between the contractor and the project's owner but also assigns rights and responsibilities to each. With these rights and responsibilities comes risk. The contractor must be aware of the contract requirements so that project risk can be minimized and managed.

4.4 WHERE IS THE RISK?

Like any construction project, the contractor's risk for a green project depends on the scope of work. The risk associated with a green project that involves constructing a building from a completed design as in a traditional design-bid-build project should be no different than a conventional project. The contractor's responsibilities on a traditional green design-bid-build project should be prescriptive and clearly defined in the bid documents (as discussed in Chapter 3). Under this scenario, the owner assumes responsibility for the performance of the building and building certification. The risk for the contractor is in completing the project on time, within budget, and in accordance with the contract documents.

When the green project is a design-build project that requires the contractor to take responsibility for design as well as construction, the risk for the contractor typically includes the design-bid-build project risk as well as the risk associated with building performance. Because most green building rating systems are based on building performance, the contractor may also assume the risk of achieving the green building certification specified in the owner's design criteria package.

4.5 MANAGING GREEN CONSTRUCTION RISK

The first step in managing construction risk is to identify and define the risks faced by the contractor on a particular project. This involves reviewing the owner's request for bid or proposal in detail to determine the contractor's responsibilities assigned and the risks associated with those responsibilities (as discussed in Chapter 3). Once identified, a plan for managing each risk can be developed using one of the following four risk management methods:

- Risk retention
- Risk reduction
- Risk transfer
- Risk avoidance

The following sections discuss each of these four risk management methods and how they might be employed to manage specific risks on a green construction project.

4.5.1 Risk Retention

The decision to retain risk on a construction project can be either conscious or unconscious on the part of the contractor. An unconscious decision to retain risk occurs when the contractor is not aware of the risks that it is assuming because it has either failed to analyze the project in detail or does not understand the implications of its assigned role or responsibilities. Passive risk retention should be avoided at all cost. One of the primary objectives of this chapter is to help the contractor identify possible green construction risks and develop strategies to effectively address them so that passive risk retention can be avoided.

A risk associated with green building construction that will be discussed throughout this chapter is the situation where the owner provides a complete design for the building system and through the contract documents also assigns the contractor with the responsibility for achieving a certain green building certification and/or level of certification based primarily on that design. If the contractor enters into a contract with the owner under these conditions, it has unconsciously retained the risk. The owner on a design-bid-build project should retain the risk of certification because the owner supplies a complete design to the contractor.

Consciously retaining certain risks after careful analysis is a viable and effective means of managing risk. Active risk retention is part of construction contracting and takes on many forms beyond just making the decision to retain or deal with a certain risk. The contractor's decision to simply retain a given risk in whole or in part is essentially self-insuring the risk. If the contractor decides to retain the entire risk, then it is self-insuring itself against possible losses associated with that risk. However, the contractor might assume a finite portion of the risk by using one of the other three methods to manage the risk.

Conscious risk retention can be illustrated using the previous example. The contractor may decide to retain the risk of achieving the owner's required green building certification based on previous experience on similar projects and a detailed analysis of the project documents. Even though the risk of achieving green building certification in this case should be the owner's, the contractor may decide that the risk of not achieving the certification is very small given the project design or that the risk is acceptable given other considerations. In consideration for retaining this risk, the contractor could increase its price to perform the work, including the anticipated additional direct costs associated with obtaining the required green certification plus allowances for unanticipated costs, overhead associated with the additional work, and profit.

4.5.2 Risk Reduction

Once identified, contractual risks can be reduced to an acceptable level through negotiation. Using the previous example to illustrate risk reduction, the contractor recognizing the risk that it is faced with can enter into negotiations with the owner to reduce this risk to an acceptable level. It may be that the owner or the architect who drafted the contract documents for the owner is not fully familiar with the certification process or aware of the additional risk that it is imposing on the contractor. During contract negotiations, the owner may agree to assume responsibility for achieving the desired green building certification based on the design, and the contractor agrees to provide the required documentation for certification as well as address specific certification requirements, such as construction waste management that is under its control. Under this scenario, risk has been reduced to an acceptable level for the contractor because the green building certification requirements for which it is responsible are under its control. Similarly, the owner has taken responsibility for the design and other certification requirements that it can control. A good construction contract assigns the responsibility to the party most able to control the risk associated with that responsibility.

4.5.3 Risk Transfer

Risk transfer involves the transfer of contractual risk that is retained by the contractor to another party via a contract. A common method of risk transfer in construction is through subcontracting, where the contractor assigns some of the contractual responsibilities it assumes in the owner-contractor agreement to specialty contractors and designers. Additionally, contracts for the supply of materials and equipment to the project with manufacturers and suppliers is also an example of risk transfer. In all cases, the contractor needs to make sure that applicable green project requirements are incorporated into subcontracts and supply agreements to ensure that the party most able to fulfill the green project requirement and control the risk associated with it is assigned the obligation through the subcontracting and procurement processes. This chapter is focused on the owner-contractor agreement for green building construction. Design agreements, subcontracting, and procurement for green construction are covered in Chapters 5, 6, and 7, respectively.

Insurance is a two-party contract in which another person or company agrees to assume the contractor's risk for specific occurrences in return for compensation usually referred to as a premium. The contract between the insured and insurer is usually referred to as an insurance policy. Contractors today carry a variety of insurance policies to protect their businesses from unanticipated occurrences both on and off the jobsite. Insurance that a contractor carries may be required (1) by law, as in the case of Worker's Compensation Insurance (WCI); (2) by contract, which may require Professional Liability Insurance on a design-build project; or (3) by choice, where the contractor chooses to carry Contractor's General Property and Equipment Insurance for its own protection.

In general, green construction does not directly impact the contractor's insurance coverage. However, just about anything can be insured, and if the contractor wanted to insure its risk associated with the green building certification in this section's example, it probably could. The premium required by the entity assuming the risk for the contractor or the terms and conditions of the insurance contract may be prohibitive, but obtaining insurance would be an example of risk transfer.

4.5.4 Risk Avoidance

The best method of risk management is to avoid it wherever possible. In order to avoid risk, the contractor must first be aware of the potential risk and then take action to avoid the risk. In the example used in this section, a simple way to avoid the risk associated with achieving a given green certification for the project based on the owner's design would be to simply decline to bid or negotiate for the project.

4.6 OWNER-CONTRACTOR AGREEMENT

The owner-contractor agreement for building contraction is often a standard form document that is developed and published by an industry organization such as the Associated General Contractors of America (AGC) or The American Institute of Architects (AIA). Owner-developed agreements are often modeled after one of these model agreements. The owner-contractor agreement may incorporate the general conditions directly into the agreement, as in the case of AGC Document No. 200 entitled *Standard Form of Agreement and General Conditions Between Owner and Contractor*, or the general conditions may be a separate stand-alone document incorporated into the owner-contractor agreement by reference, as in the case of the AIA model documents. In the AIA model documents, AIA Document A101™ entitled *Standard Form of Agreement Between Owner and Contractor* provides the model agreement between the owner and contractor, and AIA Document 201™ entitled *General Conditions of the Contract for Construction* provides the model general conditions of contract.

The owner-contractor agreement typically identifies the owner and contractor that are the contracting parties and the construction project that is the subject of the agreement. Additionally, the time to perform the work, contract sum, payment terms and conditions, and other important information is contained in the owner-contractor agreement. The project scope and requirements are defined in the owner-contractor agreement through referenced documents, which include the following on a design-bid-build project:

- Contract general conditions
- Supplementary and special conditions

- Specifications
- Drawings
- Addenda
- Other listed documents

The following sections discuss each of these documents.

4.6.1 Conditions of Contract

General Conditions. Among other things, the general conditions of the construction contract define the role and responsibilities of the parties to the construction contract, set forth procedural rules for payment and dealing with disputes and claims, and define project start and completion. To date, no standard form general condition has been modified by its sponsoring organization to address green construction. However, the contractor should thoroughly review and understand the general conditions of the contract to ensure that the owner has not included green construction requirements.

Supplemental Conditions. Supplemental conditions are normally used where the owner uses standard form general conditions such as AGC Document No. 200 or AIA Document A201™. The supplemental conditions of the construction contract modify the standard form general conditions to better meet the owner's individual needs and project requirements regarding the management and administration of the project. Administrative requirements associated with green construction may be incorporated into the supplemental conditions. This could include prerequisites for progress payments or requirements for substantial completion on a green construction project that are in addition to those encountered on a standard project. In most cases, the specific green project requirements will be found in the project specifications (as discussed in Chapter 3), but these requirements may be specifically referenced or cited in the supplemental conditions as well.

For example, a prerequisite for receiving the first progress payment on a green construction project might be the submission and acceptance of the contractor's plans for the following:

- Construction Waste Management
- Proposed Materials with Recycled Content
- Proposed Regionally Extracted, Harvested, or Recovered Materials

- Proposed Certified Wood Products
- Construction Indoor Air Quality (IAQ) Management

There may also be reports that need to be submitted with regular progress payment requests as a condition for payment. These reports could include the following:

- Construction Waste Reduction Progress
- Material Recycled Content Used
- Regionally Extracted, Harvested, or Recovered Materials Used

Substantial completion is defined in AGC Document No. 200 paragraph 2.3.17 as follows:

> Substantial Completion of the Work, or of a designated portion, occurs on the date when the Work is sufficiently complete in accordance with the Contract Documents so that the Owner may occupy or utilize the Project, or a designated portion, for the use for which it is intended.

For a traditional building construction project, substantial completion is usually achieved when (1) the contractor receives the certificate of occupancy for the building from the local authority having jurisdiction; (2) the architect has inspected the project and prepared a punchlist of minor items that need to be corrected or completed; and (3) the architect issues a certificate of substantial completion. At substantial completion, the owner takes possession of the project and becomes responsible for the building's operation; the contractor receives a prorated portion of its retainage based on the architect's punchlist; and contractually required warranties commence.

On a green project, the definition of substantial completion could be modified to include achieving a given building performance level or green building certification as a prerequisite for substantial completion. This could occur either explicitly or implicitly. Explicitly, the definition of substantial completion could be modified in the general conditions or supplemental conditions to include green building performance or certification requirements. Implicitly, the owner's or architect's interpretation of the definition of substantial completion could include meeting certain building performance requirements or achieving a specified level of green certification. Additionally, an outside entity such as a local government that has enacted a statute requiring that buildings achieve a certain performance level or green certification could also delay substantial completion. Any delay in substantial completion can be very expensive for the contractor and can include the cost of maintaining the building, additional work required to attain the required performance or certification, interest on payments being retained by the owner, extended warranties on equipment, extended site and home office overhead, and owner liquidated or actual damages resulting from the delay in substantial completion, among other costs.

Special Conditions. Special conditions deal with the owner's special requirements and restrictions for the project work. Special conditions address such things as the required sequencing of work, restrictions on personnel and construction operations, mobilization and demobilization, material delivery and handling, and site access and security. The general and supplemental conditions usually impact the pricing of the bid, because these documents define the risk to which the contractor is subjecting itself in undertaking the project. In contrast, the special conditions usually have a direct impact on the estimated cost of performing the work and often impact schedule and productivity. The contractor needs to review the special conditions of the contract carefully to identify any green requirements that could impact the cost or time to perform the work.

An example of a special condition that could impact the contractor's performance of the work is partial occupancy of the building before overall substantial completion. Partial occupancy of any building project will usually impact the contractor's sequencing of work, schedule, and productivity and needs to be accounted for when bidding the project. However, on green projects, partial occupancy may also require a separate indoor air-quality (IAQ) management plan for occupants during construction, which goes beyond that required for the construction IAQ management plan. Maintaining IAQ in areas occupied by the owner during construction may require adjusting the HVAC system installation schedule and sequence, as well as temporarily or permanently modifying the HVAC system. Early startup of the HVAC system may also result in additional costs associated with extended warranties from HVAC equipment manufacturers to meet the contractor's contractually required warranty period for the building, cleaning of ductwork, replacement or filters, repairs and maintenance, among other costs.

Combined Supplemental and Special Conditions. On most building projects, the supplemental conditions are combined with the special conditions into one document instead of providing two separate documents. The combined document is usually just referred to as either the supplemental or special conditions and includes modifications, clarifications, and additions to the standard set of general conditions used as well as the special requirements for construction.

4.6.2 Drawings and Specifications

Green construction requirements can be found anywhere in the contract documents, including the drawings (as noted in Chapter 3). With the exception of small projects, where the project specifications are included in the drawing set, green requirements will typically not be included on the drawings. However, the contractor needs to review the drawings in detail to ensure that there are no explicit green project requirements or references to green project requirements on the drawings.

Green project requirements are usually found in the specifications rather than the other contract documents, because green construction is about the materials and equipment incorporated into the construction of a building as well as the overall performance of building equipment and systems. Green building projects are also concerned with sustainable construction practices that are normally addressed in the project specifications as well as building commissioning and closeout. Specifications are an extremely important part of the contract documents on green construction projects and must be thoroughly analyzed to accurately determine the scope of work including green construction requirements, assess risk, estimate project costs, and plan the work on a green building project.

Just because the green requirements for the installation are not shown or referenced on the drawings does not mean that the contractor is not required to meet the specification requirements. Both the project drawings and specifications are contract documents by reference, as illustrated by the definition of contract documents provided in Paragraph 2.3.4 of AGC Document No. 200 entitled *Standard Form of Agreement Between Owner and Contractor*:

> The Contract Documents consist of this Agreement, the drawings, specifications, addenda issued prior to execution of this Agreement, approved submittals, information furnished by the Owner under Paragraph 4.3 [Worksite Information], other documents listed in this Agreement and any modifications issued after execution.

Furthermore, the drawings and specifications are intended to be complementary, and there is no intent to repeat all installation requirements in both documents. This is illustrated in Paragraph 14.2.1 of AGC Document No. 200, which addresses the intent of the contract documents as follows:

> The drawings and specifications are complementary. If work is shown only on one but not on the other, the contractor shall perform the Work as though fully described on both consistent with the Contract Documents and reasonably inferable from them as being necessary to produce the intended results.

In addition, many construction contracts contain an "order of precedence," which lists the hierarchy of contract documents in the event of a conflict among them. In most cases, the specifications take precedence over the drawings, as is discussed in Section 4.12.

4.6.3 Addenda

Contract conditions, plans, and specifications are never perfect. During the bid period, questions and issues surface as a result of the owner review, designer checking, or contractor analysis. When problems are discovered, the owner or designer devises a solution, and the contractor is notified of the change through an addendum. Addenda can significantly add, delete, or change the character of the work. The contractor preparing a bid for a green building project should keep abreast of all addenda issued during the bid period to ensure that all requirements are addressed in its bid. Most standard bid forms require that the contractor list all addenda received during the bid period and incorporated into the bid price.

4.6.4 Other Listed Documents

In addition to the documents that are typically incorporated into a traditional design-bid-build construction contract, the owner may incorporate others by reference that can impact the contractor's scope of work, planned means and methods, sequencing and scheduling, and productivity. Documents included by reference could be anything, including an outline project schedule that includes owner-required milestones and work sequencing, information regarding the owner-furnished materials and equipment, or test and performance reports, among others. On green construction projects, requirements such as a commissioning plan developed by a commissioning agent under contract with the owner, building performance or green building certification criteria, and other documents might be included in the contract by reference. The contractor needs to carefully review the list of reference documents and be fully aware of their requirements.

4.7 GREEN DESIGN-BUILD REQUIREMENTS

4.7.1 Design-Build Project Delivery

Green construction is not limited to the traditional design-bid-build project delivery system. Green construction can be accomplished under any project delivery system, including design-build that continues to grow in popularity in commercial and institutional building construction. Design-build offers several advantages to the owner, including having only one contract with the design-builder, who provides single-point responsibility for the project. When the contractor assumes the role of design-builder, which is common in commercial and institutional design-build projects, it usually contracts with an architect to provide design services. Managing design on a green design-build project as well as the design builder–architect agreement are covered in Chapter 5.

4.7.2 Contract Documents

The documents included in a design-build contract are not as consistent from contract to contract as they are on traditional design-bid-build projects. Design-build contract documents typically include the owner–design builder agreement; general, supplemental, and special conditions; and any amendments prior to the submission of the bid or proposal, just like a traditional design-bid-build project. The drawings and specifications are replaced by the owner's project criteria on a design-build project, which can be anything from an itemized list of performance criteria for the building project to 30 to 60 percent drawings and specifications in the case of design-assist or draw-build projects. Other documents that may be included by reference in a design-build contract include the design-builder's project proposal or bid, the design-builder's list of deviations from the owner's project criteria, or the owner's request for proposal or bid, among others.

4.7.3 Design-Build Risk

On a design-build project, the contractor assuming the role of design-builder is responsible for designing and constructing the project in accordance with the criteria included in the design-build agreement. On a green design-build project, the owner's design-build criteria may include required building or building system performance, required green building certification, or both. Because the contractor as design-builder is responsible for delivering a completed project in accordance with the owner's stated criteria, which includes the design of the building, the contractor can be responsible for the required building performance and/or green building certification. Therefore, on a design-build project, the contractor often assumes both the risk associated with building construction as on a traditional design-bid-build project and the risk associated with meeting the owner's stated building and building system performance requirements and green building certification.

Again, care must be taken when analyzing a design-build project and deciding whether to submit a bid or proposal for the project. If the contractor is responsible for meeting specified building or building system performance criteria or achieving a given green building certification, the contractor must carefully examine and analyze the owner's project criteria to ensure that meeting the owner's project criteria is consistent with the performance and certification goals. This is especially true when the project is a design-assist or draw-build project, where the major system decisions have already been made by the owner and its designers, and the design-builder's role is limited to completing the design and installation.

4.7.4 Establish Measurable Design-Build Performance Requirements

One of the major risks faced by any design-builder on a design-build project is poorly defined performance criteria. Performance criteria on a traditional design-build project should be both clearly defined and measurable. If performance criteria are not well-defined, the design-builder should define them in its proposal or bid. Similarly, if performance criteria are not measurable, the design-builder should include appropriate metrics based on industry standards in its proposal or bid. These measurable performance requirements should not only serve as the basis for project design but also for documenting and executing the building commissioning plan.

For green design-build projects, an objective industry standard for establishing system performance criteria or certification requirements should be adopted and included in the contract documents. In addition to green building certification, the design-build contract should also specify the level of certification that the building will achieve so that there are no misunderstandings. Point-based green building certification systems such as USGBC's LEED™ *Green Building Rating System* or GBI's *Green Globes™ Rating System* give the design-builder flexibility as to how it achieves the necessary points for a given level of certification. The design-builder should include a plan in its proposal delineating exactly how it will earn the necessary points for certification. This will avoid any misunderstandings between the owner and design-builder as to how certification will be achieved. For example, the design-builder may put more emphasis on green construction practices and the use of recycled materials to achieve certification, whereas the owner may have preferred a focus on building energy efficiency that would reduce its operation costs over the life of the building.

4.8 ORDER OF CONTRACT DOCUMENT PRECEDENCE

All of the documents that comprise a construction contract are intended to be complementary. However, requirements often conflict among contract documents, and there needs to be a way to resolve this conflict. This is done in most construction contracts by establishing an order of precedence among contract documents. This order of precedence establishes a hierarchy among contract documents that can be very helpful in interpreting or applying the contract. This order of precedence is important on green construction projects because the additional requirements imposed on materials and equipment in the specifications may be at odds with requirements contained in other contract documents. For example, the drawings may contain a detail that cannot be built using recycled wood, which is not graded and required to be used by the specifications.

During the review process, the contractor should first determine if an order of precedence is included in the contract documents. For example, AGC Document No. 200 has a stated order of precedence for contract documents, whereas AIA Document A101™ and its associated AIA A201™ general conditions do not include an order of precedence. If a conflict exists between the drawings and specifications, Paragraph 14.2.2 of AGC Document No. 200 states that the specifications take precedence over the drawings as follows:

> In case of conflicts between drawings and specifications, the specifications shall govern.

If there is an order of precedence, the contractor should be aware of the order and its importance in interpreting the construction contract.

4.9 INSURANCE COVERAGE ON GREEN PROJECTS

Insurance coverage required on a green construction project should be the same as on standard construction projects. However, the contractor should always review contract requirements in detail to ensure that it has the contractually required insurance. When in doubt, the contractor should review the contact insurance requirements with its insurance carrier to ensure that the owner's insurance requirements are met.

4.10 BONDING REQUIREMENTS ON GREEN PROJECTS

Bonding, like insurance, should be no different on a green construction project than on a traditional construction project. The exception to this may be the performance bond in the situation where the contractor takes responsibility for achieving a certain level of building or building system performance or green building certification, which would be most common on a design-build project. Sureties typically prefer to exclude design services and system performance from performance bonds and just guarantee the contractor's performance during construction, as they do on a traditional design-bid-build project. It is very important to determine what the owner wants covered by the performance bond as well as what the surety will guarantee under the performance bond, particularly on design-build projects or projects that include promises regarding system performance or green building certification beyond the normal construction process.

4.11 CASE STUDY

Alberici Corporation

Alberici Corporate Headquarters, Missouri

Alberici Corporation adaptively reused an existing manufacturing plant to create a new headquarters building for its company. A 50-year-old structure on a brownfield site was used as the platform for the new headquarters building. Visitors are welcome to the site, where they are educated about the project through environmental graphic displays. The flexibility integrated into the design and construction of the building permits Alberici to make productive use of the space for many years to come.

Building type(s): building	Commercial office
Size:	110,000 sq.ft. (10,100 m^2)
Completion date:	December 2004
Delivery type:	Design-Build
Rating or other recognition:	LEED-NC Platinum
Globes	Four Green
Ten 2006	AIA/COTE Top
America Award 2005	AON Build
Award 2006	EPA Phoenix

Environmental Aspects

A "saw-tooth" shaped building addition was used to reorient the rear portion of the building due south. External sunscreens were added to block unwanted solar gain. The interiors are designed around three atria, which serve as thermal flues. 100 percent of building occupants enjoy direct views to the outdoors.

During construction, Alberici developed a construction compliance program that included erosion and sedimentation control, a stormwater pollution prevention plan, an indoor air quality plan, and a waste management plan to divert construction and demolition waste from landfills.

Figure 4-1 Photo Courtesy of Debbie Franke

The site was restored with native, drought-resistant plantings. Retention ponds and constructed wetlands retain all of the stormwater runoff. Rainwater is collected and used for 100 percent of sewage conveyance. The captured rainwater is also used in the mechanical system's cooling tower.

Onsite renewable sources generate nearly 20 percent of the building's required energy. A 65-kilowatt wind turbine provides 92,000 kilowatt-hours, 18 percent of the facility's electrical needs, annually. Solar panels are used to preheat hot water. The design features a lighting power density that is about 25 percent of the power typically used for lighting an office building.

Lessons Learned

- The use of a design-build delivery method was a natural fit for the project. Alberici served as the developer, design-build contractor, sustainability consultant and is the occupant of the building. Design-build provided the flexibility and the control of the process that enabled the team to push the project as far as they did.
- Education of the tradespeople working on the project was one key to the success of the project. The site management team took time during each person's orientation to teach them about sustainability and how their everyday actions would affect the outcome of the building.
- At the time of construction, many of the material suppliers did not know how to respond to the requests for information and the documentation required for the LEED certification process. The team spent multiple man-hours working with manufacturers to assemble the paperwork required by the USGBC.
- Site management on a project that is seeking certification is quite different from a typical project. The site supervision must enforce compliance with Construction Indoor Air Quality protocols as well as Construction Waste Management plans on a daily basis.
- A contractor (prime or sub) cannot rely on the traditional contract documents for all of the requirements that are imposed on a certified project. The site team must understand the contract documents and know the requirements of the rating system they are using to certify the project.

4.12 REFERENCES

The American Institute of Architects, *Standard Form of Agreement Between Owner and Contractor (Where the Basis of Payment Is a Stipulated Sum)*, AIA Document A101™-1997, 1997.

The American Institute of Architects, *General Conditions of the Contract for Construction*, AIA Document A201™-1997, 1997.

The Associated General Contractors of America, *Standard Form of Agreement and General Conditions Between Owner and Contractor (Where the Contract Price Is a Lump Sum)*, AGC Document No. 200, 2000. [Refer also to pending ConsensusDOCS: General Contracting Number 200 entitled *Owner/Contractor Agreement & General Conditions (Lump Sum)*, 2007.]

chapter 5

Managing Green Design

5.1 INTRODUCTION

The use of design-build as a project delivery system by owners is growing in the United States for commercial and institutional facilities. As a result, the contractor that is assuming the role of design-builder on a green project needs to understand how to manage not only construction but also the design process. Green building design can be more complex than what is usually required for a traditional building. Sustainable design typically requires the evaluation of alternative materials and systems by the design-build team. Building modeling and simulation to ensure that the constructed building will meet the owner's performance requirements and conform to industry standards, energy codes, and rating systems are inherent on a green building project. This chapter addresses managing green design for design-build projects. This chapter should also be useful to the construction manager in working with the owner and designer during the planning and design process for a green project.

5.2 DESIGNER DEFINED

The term *designer* is used throughout this chapter to identify the entity that is responsible for the project design. The designer could be an individual, but, except for the smallest projects, the designer referred to in this chapter is a firm or more likely a group of firms with the necessary expertise and qualifications to design the project. The designer includes an architect or architecture firm that would normally be the lead designer and architect of record on a building project. Supporting the architect would be individual engineers or engineering firms with expertise in the design of structural; heating, ventilating, and air-conditioning (HVAC); plumbing and fire protection; power and communications; building controls; and other engineered building systems. In addition, the designer could include individuals and firms with expertise in interior design, acoustics, artificial lighting and daylighting, security and life safety, and other specialized design areas.

5.3 DESIGN-BUILD AS A PROJECT DELIVERY SYSTEM

Design-build is the project delivery system where one organization, referred to as the design-builder, is responsible for both project design and construction. Under this project delivery system, the owner has only one contract for the complete delivery of the project, including both design and construction. Design-build reduces the number of contracts needed by the owner for a building project. In addition, because both the contractor and designer are part of the same entity, communication and interaction during the design and construction process should be improved, which is important given the increased complexity of a high-performance building project.

5.4 DESIGN-BUILDER DEFINED

The entity that the owner contracts with is referred to as the design-builder. *Design-builder* is defined by the Design-Build Institute of America (DBIA) as follows (DBIA 1996):

> The entity contractually responsible for delivering the project design and construction. The design-builder can assume several organizational structures: the firm possessing both design and construction resources in-house, a joint venture between designer and contractor, a contractor-led team with the designer in a subcontract role, or a designer-led team with the constructor in a subcontractor role.

As noted in the definition, the design-builder can assume any of the following organizational structures:

- Firm with in-house design and construction
- Joint venture between the designer and constructor

- Contractor-led team with designer as subcontractor
- Designer-led team with contractor as subcontractor

No matter how the design-builder is organized, the only thing that the owner sees contractually is a single entity. To the owner, it does not matter how the design-builder is organized. However, the organization of the design-builder can impact the efficiency and effectiveness of project delivery and should be tailored to the specific project needs. Any one of the four ways to organize as a design-builder will work, and the design-builder can choose the organizational model that best fits the circumstances of a particular project. However, the most prevalent organizational model for commercial and institutional building design-build projects is the contractor-led team with the designer as a subcontractor. Therefore, this chapter focuses on this organizational model and the unique challenges faced by the contactor on green design-build projects. Throughout the remainder of this chapter, the design-builder will be referred to as the contractor.

5.5 DESIGN-BUILD IS A TEAM EFFORT

5.5.1 Relationship between Contractor and Designer

Design-build is a team effort where both the contractor and designer work together toward the common goal of providing the owner with a facility that meets its needs in a timely and cost-effective manner. This is especially true for green design-build projects, where the design is instrumental in meeting the project's sustainability goals. However, in this chapter, the contractor is assumed to be the design-builder and has a direct contract with the owner, and the designer is a subcontractor to the contractor. As a result of this contractual arrangement, the contractor is responsible for designing the project to meet the owner's project criteria. Therefore, throughout this chapter, the contractor is noted as the responsible party for the performance of the design to the owner, even though the designer will perform the design through its subcontract with the contractor.

5.5.2 Designer Organization

As noted previously, the designer on a design-build project usually consists of a group of architects, engineers, and other design professionals. The way in which the designer is organized will vary from project to project. An architect-led design team is common on building projects where the architect contracts directly with the contractor and then subcontracts specialty design work that it cannot perform in-house to outside engineers and designers. Alternatively, with a distributed design organization, the architect could contract directly with the contractor for its work, and needed specialty design could be performed by specialty contractors with the in-house design capabilities or subcontracted to a specialty designer by the specialty contractor. For example, under this scenario, the electrical contractor would be responsible for the design of the building power distribution system either by performing the design in-house or subcontracting the design to a consulting engineer.

This distributed approach to design can work well for traditional buildings where building systems are still designed and operated as stand-alone systems. However, the distributed designer organization can be cumbersome and even a hindrance on green design-build projects, because high-performance buildings require that building systems be integrated and interoperable. This means that the design team must also be integrated and have a very high degree of interaction throughout the design process, which will be much more difficult under the distributed model. The designer organization is very important on a green design-build project and needs to promote teamwork, interaction, and communication.

5.6 WHAT MAKES A SUCCESSFUL DESIGN-BUILD PROJECT?

A successful design-build project is one that is:

- Completed on time
- Completed within budget
- Meets the owner's needs and expectations

These same three criteria make any construction project successful, no matter what project delivery system is used. However, the difference is that with the design-bid-build and construction manager project delivery systems, the owner's needs and expectations are determined by the designer and documented in the plans and specifications. If the contractor performs the work in accordance with the plans and specifications, then it has typically fulfilled its contractual obligation. Any problem with building performance is the designer's responsibility. If there is a problem with the plans and specifications or if the building does not perform properly after construction, the contractor should not be responsible if the building is built and systems are installed in accordance with the contract documents. If the contractor is involved in fixing the building performance, then the contractor typically receives a change order than can include additional compensation and time.

With design-build, however, the design-builder assumes the risk for providing a building that meets the owner's needs and expectations as expressed in the contract documents. For a traditional building project, owner requirements and expectations often focus on the physical layout and construction of the building, and building operational characteristics are secondary. For green buildings this priority is reversed, where building operational characteristics are the primary focus of the design and construction process, and these factors impact the physical layout and construction of the building.

This change in focus on a green design-build project requires a goal-oriented approach to design that is based on measurable building performance criteria. The design process is much more involved and important on a green design-build project. A lot of design and construction overlap that can occur on a traditional design-build project cannot occur on a green design-build project. As a result, the design process will typically take longer for a green design-build project than for a traditional design-build project.

Green buildings require a holistic and integrated approach to design that precludes treating the building layout, shell, and systems separately. Instead, the building needs to be designed and performance optimized as an integrated system and not a group of independent systems, as is typical for traditional buildings. The contractor operating as a design-builder on a green construction project needs to be aware that design will take longer and will need to be more complete before starting construction. This will impact the project schedule, increase design costs, and require greater coordination, and building operational performance will be as or more important than the physical structure. Building performance is the biggest risk faced by the contractor on a green design-build project.

5.7 UNDERSTANDING OWNER NEEDS AND EXPECTATIONS

5.7.1 Owner Needs and Expectations

The owner's needs and expectations on a traditional construction project where the design is completed by the designer under contract with the owner prior to negotiating or bidding the project construction are expressed in the bid documents, which include the plans and specifications. As noted previously, the contractor's obligation on a traditional project is limited to construction of the project in accordance with the contract documents and, except for specific portions of the project that may be detailed using a performance specification, the contractor is not responsible for the building's performance.

For a design-build project, the owner's needs and expectations are specified in terms or criteria that must be met, which may include not only prescriptive criteria such as that the façade should be a certain style or to use a certain material to blend with surrounding buildings but also requirements for building system performance. With green design-build projects, the design-builder may also be faced with other criteria, such as building operational criteria, life-cycle issues, an outside commissioning agent working directly for the owner, or a specific certification of green building status by an outside third-party organization. The design-builder must thoroughly understand the project's green requirements.

5.7.2 Owner's Project Criteria

The owner's request for proposal (RFP) for a design-build project will include the project criteria that are developed by the owner or with the help of an outside criteria consultant. The owner's project criteria can be anything from a detailed and comprehensive package geared to the specific project to a general description of the project, leaving it to the design-builder to fill in the details. In either case, the owner's project criteria—sometimes referred to as either the design criteria package or statement of facility requirements—is the part of the RFP that defines the owner's needs and expectations for the physical building project. The owner's project criteria is defined by the DBIA as follows (DBIA 1998):

> Owner's project criteria are developed by or for the owner to describe the owner's program requirements and objectives for the project, including use, space, price, time, site and expandability requirements, as well as submittal requirements and other requirements governing design-builder's performance of the work. Owner's project criteria may include conceptual documents, design criteria, performance requirements and other project-specific technical materials and requirements.

The design criteria package should define the specific green requirements that the design-builder is expected to meet, such as reduced site disturbance, optimized energy performance, and incorporation of daylighting and lighting control, among many other criteria. Alternatively, the owner's criteria could simply state that the design-builder achieve a certain green building status as certified by an outside third-party organization. No matter how it is stated, the design-builder needs to understand the owner's criteria and develop a plan for meeting it.

5.7.3 Owner's Project Criteria as a Contract Document

The owner's project criteria document typically becomes a part of the agreement between the owner and the design-builder by reference on design-build projects. Furthermore, depending on the contract, the owner's project criteria may take precedence over the contractor's approved design, making the owner's project criteria the controlling criteria in the event of a conflict between the contractor's approved design and the owner's project criteria. The incorporation of the owner's project criteria document is an important consideration on any design-build project, but it is especially important on a green building project, where one of the major risks faced by the design-builder is building performance.

It is very important that green building performance criteria be measurable and not stated vaguely, because the owner's project criteria should serve not only as the basis of the design but also as a tool to evaluate building performance after construction. Equally important to including the owner's project criteria in the contract documents is including a document by the design-builder stating how each of the owner's requirements will be fulfilled. This is one way of measuring deliverables to stated objectives.

5.7.4 Importance of Measurable Green Criteria

The contractor should always make sure that the owner's green project criteria are measurable and achievable. For example, the owner's project criteria may require that the building use 15 percent less energy than a comparable conventional building. This requirement could be interpreted and met by the contractor in a variety of ways. Is the reduction based on actual energy units such as kilowatt-hours (kWh) of electricity or British thermal units (Btu) of natural gas, or is the specified reduction to be based on the reduction of building's overall energy expense? What constitutes a comparable building? Is the specified energy reduction to be achieved by the building as a whole or by specified systems? What is the benchmark for comparison, and how do energy codes and standards impact this requirement? If the owner requires the incorporation of photovoltaics (PV) in its project criteria, can the PV energy production be counted toward meeting the energy reduction criteria? These questions illustrate how a simple requirement like reducing building energy use by 15 percent can be interpreted in many ways by the contractor.

In the case where the design-builder creates a document that aligns the owner's expectations to the design-builder's deliverables, this requirement could be clarified as follows by the design-builder:

> The building will reach a 15 percent increase of efficiency compared to the same building constructed to meet ASHRAE 90.1-2004 as a minimum standard and will be modeled using the Department of Energy DOE2 energy model or its equivalent.

Requirements without specific definition could result in further confusion if the owner's intent is to use this 15 percent energy-use reduction to fulfill a specific green building requirement or rating system component that has specific measurement criteria associated with it and the owner's intent is not communicated to the contractor. Green criteria must be well-defined and measurable to avoid misunderstandings and disagreements at the end of the project that could mar an otherwise successful project.

Sometimes owners do not establish detailed performance criteria for the design-builder to follow, especially in negotiated work where the design-builder is the only firm being considered by the owner. If the owner does not include measurable green performance criteria in its RFP, then the design-builder needs to establish these requirements and convert them into measurable design criteria. These measurable performance criteria will determine when the work is complete and when the design-builder has fulfilled its contractual obligations to the owner. This is why the establishment of measurable design criteria is so important.

When no performance criteria is explicitly provided or it is not detailed, there is always the tendency to want to avoid establishing criteria out of fear of not being able to meet the self-imposed criteria and being held to them.

This is a risk, but the contractor should know that the bigger risk is the potential dispute between it and the owner at the end of the project when the building does not meet the owner's expectations, even though they were never fully expressed. Incorporating measurable green performance criteria in the design-build proposal will help eliminate misunderstandings and disputes at the end of the project.

5.7.5 Establishing Measurable Green Criteria

Establishing measurable green criteria that evolves into system performance criteria should start with the owner's RFP. If the owner provides a detailed RFP, then much of the work of defining green performance criteria may have already been done by the owner, and all the design-builder has to do is verify that the design criteria is complete, reasonable, and measurable. If not, then the design-builder needs to establish measurable green criteria that will be the basis for both building design and commissioning. If the design-builder establishes the criteria, then this criteria needs to become part of the design-build contract.

The next step in defining measurable green criteria is to understand the owner's needs and expectations. What the owner asks for in its RFP may not be what it actually wants or needs. The contractor needs to understand exactly what the owner wants and needs and identify any discrepancies between what is explicitly stated in the RFP and implied during its analysis of owner needs. This step is also very important in identifying areas where alternate systems, equipment, and materials will provide the owner with the same function at a reduced cost. Understanding the owner's needs may not only reduce the contractor's risk but also provide it with competitive advantage if it is competing against other firms for the project.

After reviewing the RFP and gaining an understanding of the owner's actual needs, the design-builder should develop measurable green criteria for the project using existing green building rating systems and industry codes, standards, and recommended practices (discussed in Chapter 3). The design criteria must be defined in terms of the project and documented for both approval by the owner as well as use by the design team. These measurable green criteria will serve as the basis for design and later for testing to determine that the building meets the owner's stated requirements.

5.7.6 Beware of Hidden Green Requirements and Conflicts

Last but not least, the contractor on a green construction project needs to beware of hidden green requirements and conflicts. For example, many states and municipalities currently have laws and statutes or are in the process of developing them that require buildings to be designed and constructed in accordance with specified energy codes and green building standards and rating systems. On a traditional building project, these requirements are incorporated into the design by the designer, and the contractor complies by implementing the design during construction. However, on a design-build project, if the contractor operating as a design-builder enters into a design-build agreement with an owner before engaging a designer and is unaware of these implicit project requirements, the contractor could find itself in a bad position before the project even starts.

The design-builder also needs to beware of conflicts built into the owner's RFP and address them in its proposal. For example, the owner's project requirements might require waterless urinals as a way of increasing building water efficiency, but the use of waterless urinals may be prohibited by the local plumbing code. Similarly, the owner's project criteria may define building orientation and window area that could have a significant impact on the design-builder's ability to meet energy reduction requirements.

5.7.7 Basis for Commissioning Plan

The owner's project criteria converted into measurable design criteria on a design-build project should also be the basis for the contractor's commissioning plan. The contractor's commissioning plan is different than the commissioning plan required by many third-party green building rating systems to be prepared by an outside commissioning authority. The contractor's commissioning plan would take the commissioning authority's plan and break it down into specific work activities, which are then incorporated into the project schedule and assigned to the responsible specialty contractor working for the design-builder. The building commissioning plan is very important on any design-build project, because the design-builder demonstrates that it has met the owner's project criteria through the building startup and testing process. This is especially true for green building projects that could include an independent commissioning team. The commissioning plan should address green requirements and be developed in conjunction with the project design criteria.

5.8 SELECTING A DESIGNER

5.8.1 Need for a Designer

Most contractors do not have the necessary in-house design capability for a design-build project and need to retain an outside designer. The contractor needs to know how to identify potential designers, solicit proposals from them, evaluate their submittals, interview short-listed designers, and select a compatible designer for the design-build project. The selection of the designer should be based on the project scope and the designer's experience and expertise. This is especially true on green building projects, where it is imperative that the designer has a background in green design and construction as well as thoroughly understands any green rating systems being used on the project.

5.8.2 Defining the Design Firm's Scope of Services

In order to identify and select a design firm, the contractor must first define the expertise and scope of design services that it needs. This information can then be used to draft a request for qualifications (RFQ), which will be sent to potential design firms as part of the selection process. The contractor needs to develop a clear and unambiguous description of exactly what services it needs from a design firm so that it can select the best fit.

5.8.3 Identifying and Selecting a Designer

Selecting a design firm includes the following five steps:

Step 1: Identify potential designers

Step 2: Request qualification statements

Step 3: Prepare a short list

Step 4: Interview designers

Step 5: Select the preferred designer

Step 1: Identify Potential Designers. The first step in selecting a design firm is to identify potential firms, in any of the following ways:

- Past experience
- Recommendation of other contractors
- Professional organizations
- Published listings such as the Yellow Pages

One important consideration in identifying a designer for a green building project is that designer's experience and expertise in green design. Another important consideration is the designer's familiarity with the third-party green building rating system that will be used on the project. It is important to note that, while growing rapidly, the green building movement is still very small. Many designers claim experience with green building projects but have had little or no experience with the third-party certification process or rating system being used. It is very important that the contractor understand the designer's experience not only in green design but also the third-party certification process and rating system being used.

Step 2: Request Qualification Statements. Once potential design firms have been identified, the next step is to request qualification statements from each of them. The purpose of requesting qualification statements is to determine if the firm has the necessary expertise, experience, and resources to perform the work. Requesting qualification statements from design firms normally requires that the contractor do the following:

- Prepare a description of the project.
- Contact potential firms to determine if they are interested in the project.
- Send each interested firm an RFQ that defines specifically the information needed.
- Set a deadline for return of the firm's qualifications.

Step 3: Prepare a Short List. Once received, the contractor should review the qualification statements submitted by the designers to determine the following:

- Does the designer have the necessary expertise, experience, and resources to perform the work?

- Does the designer have a history of completing projects on time, within budget, and meeting client needs? This can be verified by requesting references and following up with them.
- Does the designer have and use compatible software and information technology (IT) systems?

While often considered synonymous, expertise and experience are two different things when it comes to evaluating designer qualifications. Expertise refers to the designer's current green design capabilities, whereas experience refers to its past work on green projects. It may well be that a design firm has a great deal of experience in green project design as represented by its past project list but no longer has the in-house expertise because the designers who worked on these green projects left the firm to form their own design firm or joined other established design firms. Based on its review of the qualification statements received, the contractor should prepare a short list of acceptable designers to interview.

The need for the designer to have and use compatible software and IT systems is very important on green design-build projects where the contractor and designer have to work as a team throughout the design and construction process. Because of the iterative nature of green design, as well as the modeling, simulation, and documentation requirements included either implicitly or explicitly in the owner's project criteria and third-party certification and rating systems, having a common software platform for exchanging information will greatly improve communications, enhance design quality and efficiency, and reduce both project time and cost. Building information modeling (BIM) can be used to model and simulate the construction and operation of the building under design, making it easy for the designer and contractor to analyze different options and perform what-if analyses.

A five-dimensional model would include a three-dimensional model of the building plus time sequence and material quantities and costs. For the designer, BIM facilitates the design process by letting it quickly make changes and build models that allow it to explore alternatives from the standpoint of building use and operation. From a construction standpoint, BIM can be used by the contractor to [AGC BIM 2006]:

- Easily identify collisions such as ductwork colliding with the structural steel frame of the building.
- Visualize what is to be built in a simulated environment.

- Better anticipate actual field conditions.
- Perform what-if scenarios such as investigating different sequencing options, site logistics, and hoisting alternatives, among others.

Step 4: Interview Designers. Interviewing potential design firms before selecting the firm to work with is very important. This step is probably the most crucial step in selecting a designer. Based on qualifications statements, the contractor has a set of technically qualified designers to perform the work. This step should determine if the contractor is able to work with the designer. Not all designers are able to work effectively for a contractor firm in a design-build mode.

Step 5: Select the Preferred Designer. Based on all of the information gathered, select the designer firm with which you are most comfortable. Remember, you are selecting a project partner.

5.8.4 Professional Qualifications

Professional qualifications on green design-build projects for the design firm and designer include both professional licensing and green qualifications. The following paragraphs will discuss each of these professional qualifications for green design-build projects.

Professional Licensing. The designer performing the work or the principal of the design firm responsible for performing the design needs to be a registered design professional in the state where the project is being built. In many jurisdictions, the authority having jurisdiction will require that a registered architect seal the architectural drawings and specifications and a registered engineer from each discipline seal the respective engineering drawings and specifications.

Green Qualifications. In addition to professional licensing, designers should be knowledgeable about green design and sustainable design strategies. Where appropriate, designers should demonstrate their green design expertise through a third-party certification or accreditation process. For example, the designer on a green building project using the U.S. Green Building Council's (USGBC) Leadership in Energy and Environmental Design (LEED) green building rating system should be a LEED-Accredited Professional (LEED AP), which involves successfully passing the LEED accreditation examination. In addition to demonstrating expertise, having a lead designer who is a LEED AP will streamline the green certification process for the building project and earn one point toward LEED accreditation.

5.9 CONTRACTING WITH THE DESIGNER

5.9.1 Design Contract Scope

Once the preferred designer has been selected, the next step is to contract with the designer to perform the needed design work. Contracting with the designer requires a written agreement that outlines the roles and responsibilities of each party for completing the work, along with other important provisions, including payment, insurance, and indemnification. A detailed scope of services for which the designer is responsible needs to be incorporated into the agreement either by reference or directly. This scope of services should address the green requirements that are part of the contractor's design-build agreement with the owner as discussed previously.

5.9.2 Design Contract Requirements

Among other things, the agreement for design services between the design-builder and designer should do the following:

- Define roles and responsibilities of both parties throughout the project, including design, installation, and commissioning.
- Define roles and responsibilities for third-party green building certification if it is being sought by the owner.
- Establish a method of addressing design modifications and changes that result from design reviews, value analyses, constructability reviews, owner changes, and design errors and omissions.
- Address who owns the design.
- Incorporate noncompete or nondisclosure clauses where appropriate to facilitate the exchange of information.
- Establish payment terms and conditions.
- Define insurance requirements and responsibilities.
- Provide an indemnification clause.
- Define rights and responsibilities for the termination of the agreement.
- State other terms and conditions needed for a particular project, required by your firm, or recommended by your firm's attorney, insurance carrier, and surety.

5.9.3 Design Contract Format

The agreement between the contractor and the designer can be anything from a simple letter agreement to a detailed formal contract for design services, depending on the size, scope, and complexity of the project. No matter what is used for the design contract, it must be complete and should be reviewed by the contractor's attorney, insurance carrier, and surety before execution. It is important to note that the contractor's attorney, insurance carrier, and surety may have limited knowledge of green design and the requirements of third-party green building rating systems. These individuals should take the time to educate themselves about green construction before making their recommendations.

Model contract forms for providing design services to the contractor on design-build projects can be used directly or as a good starting point for developing a custom form agreement by the contractor's attorney. These model design agreements are sponsored by industry associations, have evolved over the years, been reviewed by the industry and tested in the courts, and contain all of the elements discussed previously. However, it is important to note that none of the model design agreements was designed to address the unique aspects of green building design and construction. This means that any model design agreement used on a specific green building project will probably need to be modified. Three model design agreements developed specifically for design-build projects are as follows:

- Associated General Contractors of America (AGC), *Standard Form of Design-Build Agreement and General Conditions between Owner and Design-Builder*, AGC Document No. 410, 1999. [Refer also to pending ConsensusDOCS: General Contracting Number 420 entitled *Design-builder & Architect/Engineer Agreement*, 2007.]
- American Institute of Architects (AIA), AIA Document B901, Standard Form of Agreements Between Design/Builder and Architect (Part 1: Preliminary Design and Part 2: Final Design)
- Design-Build Institute of America (DBIA), DBIA Document 540, Standard Form of Agreement Between Design-Builder and Designer

5.9.4 Use of Purchase Orders

Many contractors use purchase orders to contract for design services. Purchase orders are used because most contractors' procurement and accounting systems are built around the purchase order. However, purchase order terms and conditions are based on the Uniform Commercial Code (UCC) and designed for procuring materials and equipment and not design services. Standard purchase orders should not be used to contract for design services.

A purchase order form can, however, be designed to procure design services that will provide the purchase order number and other information needed by the contractor's accounting system. This design services purchase order should include terms and conditions specifically developed for design services by the contractor's attorney. A design services purchase order can refer to the full agreement between the designer and the contractor and provide an easy and efficient way for the contractor to contract with the designer on an ongoing basis without executing the entire agreement each time.

5.9.5 Alignment between Agreements

The contractor's attorney should also review the contractor's agreement with the owner before finalizing the agreement between the contractor and designer for design services. Construction contracts typically attempt to develop a chain of coordinated contracts that span from the owner to the lowest-tier subcontractor or supplier using flow-through clauses. As discussed previously, these clauses often obligate subcontractors to the applicable terms and conditions of the contract between the owner and the prime contractor. Because the owner has no control over the terms and conditions between the contractor and its designer, the contract between the contractor and the owner may specify design-related insurance coverage, indemnification, and other requirements. If the contractor agrees to meet these design-related terms and conditions in its contract, it needs to incorporate these requirements into its agreement with the designer either explicitly or by reference.

5.10 PROFESSIONAL LIABILITY INSURANCE

5.10.1 Professional Liability Insurance Overview

Professional liability—or what is often referred to as errors and omissions or E&O—insurance addresses problems that can be traced back to the design of the building or installed system. Professional liability insurance protects against liability arising from errors, omissions, and negligent acts during design.

5.10.2 Need for Professional Liability Insurance

There is no legal requirement that design professionals carry professional liability insurance. Providing design services without professional liability insurance is a business decision and is often referred to as "going bare." Without professional liability insurance, the designer is putting itself at risk not only for the monetary amount of a judgment against it but also for legal fees incurred while defending itself, whether or not it is found negligent in the performance of its work.

The contractor needs to understand the risk it may be assuming if it does not require the designer to carry professional liability insurance. Because the designer is a subcontractor to the contractor, the contractor will probably be involved in any legal action resulting from the design firm's performance, whether or not professional liability insurance is required by the contractor's contract with the owner. However, if the agreement that the contractor has with the owner requires professional liability insurance for any design performed, and the contractor does not require its designer to have it, then it may be liable to the owner for any errors or omissions resulting from the designer's work.

Subcontracting the design work to a design firm may imply that the contractor is providing design services as well as its usual procurement and installation services. Professional liability for design services is typically excluded from the contractor's general liability insurance, but it is possible in some instances to get a professional liability endorsement for a project and cover an uninsured designer. Additionally, performance bonds often exclude design services or the impact of design issues on the contractor's project performance if it is providing the design services. Therefore, the contractor should work closely with its insurance carrier and surety when subcontracting to a designer on a design-build project to ensure that coverage is coordinated.

5.10.3 Insurance Coverage Period

The coverage period for the designer's professional liability insurance is a very important to the contractor acting as a design builder on a green building project. The contractor should first determine if there is a specific coverage period requirement for professional liability insurance in its design-build agreement with the owner, then specify the coverage period that it wants the designer to have in its request for proposal, and finally incorporate the required coverage period in its agreement with the designer. The following paragraphs will define what the insurance coverage period is and describe the different types of coverage periods often encountered in professional liability insurance.

Coverage Period. Every insurance policy has a coverage period that determines the time limits on the policy's coverage. This coverage period will determine whether the insurance policy is a claims-made or occurrence insurance policy. The contractor should always review the terms and conditions of the design firm's professional liability insurance policy to understand coverage periods.

Claims-Made Coverage. Claims-made insurance policies insure against claims made during the policy period. To be protected by a claims-made insurance policy, the professional liability policy must be continuously in force from the time of the occurrence to the time of the claim. Claims-made insurance policies can be problematic for construction, because to maintain coverage after construction is completed, the policy must be maintained and kept in force.

Occurrence Coverage. Occurrence insurance policies cover claims in which the insured event occurred during the policy period. With occurrence coverage, the right to protection under the insurance policy is determined by when the act or omission that caused the damage occurs. The insurance policy must have been in force at the time of the occurrence, but the policy does not have to be in effect when the claim is made. For coverage under an occurrence policy, it does not matter whether the policy is still in force when the claim is filed. An occurrence insurance policy that is in place during the project remains effective after the project has been completed for any claims that are covered and occurred during project performance.

Hybrid Coverage. There are also hybrid policies that include characteristics of both the claims-made and occurrence policies. The policy period for a hybrid insurance policy sometimes requires that the occurrence take place during the policy period and that the claim be made either during the policy period or within a specified period after the expiration of the insurance coverage.

Limitation Periods. Limitation periods also restrict insurance policy coverage. These limitation periods may be statutory or part of the terms and conditions of the insurance policy. These limitations establish time requirements for a notice of claim to be made or the initiation of a legal action that may be covered by the insurance policy.

5.10.4 Example Contractual Requirement for Professional Liability Insurance

The following is an example clause that the contractor may want to include in its agreement with the designer. This clause is the requirement for professional liability insurance contained in Section 11.3 of AGC Document No. 410 entitled *Standard Form of Design-Build Agreement and General Conditions Between Owner and Prime Contractor* [AGC 1999]:

> The Engineer shall obtain professional liability insurance for claims arising from the negligent performance of professional services under this Agreement, which shall be General Office Coverage/Project Specific Professional Liability Insurance [cross out one], written for not less than $_____ per claim and in the aggregate with a deductible not to exceed $_____. The Professional Liability Insurance shall include prior acts coverage sufficient to cover all services rendered by the Engineer. This coverage shall be continued in effect for _____ year(s) after the date of Substantial Completion.

5.11 THE DESIGN PROCESS

5.11.1 Six Design Phases

The actual design process varies from project to project depending on project participants, size, and complexity. However, on a traditional building project, the design process is usually divided into the following five phases:

- Programming
- Schematic design
- Design development
- Construction documents
- Construction administration

All design progresses through these five phases, whether it is performed by a single designer on the back of an envelope or by a team of design firms preparing the construction documents for a major construction project. The only difference may be whether the stages are explicitly identified and used for design milestones with deliverables and design reviews at the end of each phase or whether it is a continuous process that moves from programming to construction. In either case, understanding the design process in terms of these five phases will help the contractor firm better manage the design process.

On a green building project that will be certified or verified by an outside third party, an additional phase is usually required. The sixth design phase on a green building project is the commissioning phase, which follows construction administration. The following sections describe each of these six phases.

Programming Phase. In the programming phase, the contractor, usually through its designer, defines the owner's needs and expectations. The designer then translates these needs and expectations into measurable performance objectives that will serve as the basis for design. At the end of this stage, the contractor should review the program and develop a conceptual cost estimate and schedule based on past experience with similar projects and industry information. This information should then be submitted to the owner for review and approval.

Schematic Design Phase. During the schematic design phase, the designer performs code reviews and any studies or testing required for the design. The designer selects materials, equipment, and systems that will be used in the design and develops outline plans and specifications for the project in accordance with the program requirements. The contractor performs a design review that includes a constructability review, value analyses, and life-cycle cost assessments as required. After completing the design review, the contractor updates the cost estimate and schedule. The schematic design, along with the updated cost estimate and schedule, should then be submitted to the owner for review and approval.

Design Development Phase. In this phase, the designer prepares more-detailed plans and specifications that further define the project. Again, the contractor performs a design review that includes a constructability review, value analyses, and life-cycle cost assessments as required. After completing the design review, the contractor updates the cost estimate and schedule based on the increased design detail. The design, along with the updated cost estimate and schedule, is then submitted to the owner for review and approval.

Construction Documents Phase. On traditional design-bid-build and construction manager projects, the designer completes the design in sufficient detail that it can be bid out and then built. On a design-build project, the detail that the designer provides on a traditional project may not be required because the work is not open for competitive bid, and both the designer and contractor are working together as a team. The level of detail required by the contractor will depend on the complexity of the project as well as other factors, such as the abilities of personnel in the field. The extent to which the contractor needs construction documents should be determined up front so the designer can budget and schedule for them if needed.

The contractor should perform a design review at the completion of this phase before construction begins. Also, the cost estimate and schedule should be updated as they were in previous phases. The difference now is that the cost estimate will become the budget for the completion of the project, and the schedule will become the project's as-planned schedule that will be used to track project progress. The design, along with the updated cost estimate and schedule, is once again submitted to the owner for review and approval.

Construction Phase. During the construction phase, the contractor puts work in place at the project site. The designer should visit the site as required to determine that the work is being completed in accordance with the design documents. At the completion of this phase, the installation is inspected, put in service, and tested to ensure that it meets the owner's needs and expectations. During the construction phase, the owner monitors progress and, at the end of the project, inspects the work for compliance with the contract documents.

Commissioning Phase. The commissioning phase is usually included on green building projects that are required to be certified or verified as a green building using a third-party rating system. During the commissioning phase, the designer may be required to assemble documents, create drawings, perform calculations, consolidate information from other members of the design-build team, and prepare the required submission for third-party certification or verification. This can be a significant effort and needs to be addressed in both the designer's schedule and budget. Chapter 9 covers the commissioning process in detail.

5.11.2 Use of the Design Phases

As discussed previously, the design phases provide a conceptual framework for breaking down the design process and should aid in understanding and managing the design process. The breakdown can also be used to set identifiable milestones that can be used to track design progress and schedule design reviews.

The breakdown can also be helpful in deciding where the design process starts for the contractor. Depending on the owner's request for proposal (RFP), work for the contractor firm can start at any point in the process. For instance, the owner can develop a program that is the basis for its RFP using a criteria consultant, and the design-build project could start at the conceptual design phase for the contractor firm. Alternately, the owner could have its own designer, and the design could be developed through the schematic design phase, making the project essentially draw-build or design-assist for the contractor firm.

5.12 PERFORM REGULAR DESIGN REVIEWS

5.12.1 Need for Regular Design Reviews

The contractor needs to coordinate and schedule regular design reviews to ensure that the design is proceeding as planned and will meet the owner's requirements. These design reviews often occur at the end of each predefined milestone in the design process. These milestones may vary from project to project. It is important that design reviews include project personnel and are scheduled early in the design process, because design changes can be most easily and efficiently made at this time. The further into the design process, the more difficult and expensive it is to make changes.

5.12.2 Types of Design Reviews

The project manager should schedule two types of design reviews:

- Internal design reviews
- Owner design reviews

Internal Design Reviews. Internal design reviews should be scheduled by the contractor and performed at regular intervals as required by the complexity and scope of the project. The contractor should be responsible for organizing, performing, and documenting the results of these reviews. Internal design reviews should include members of the design team, field personnel, key subcontractors, suppliers and manufacturers, outside specialists, and others that are impacted by the design process. Internal design reviews should include not only a technical review of the design but also a review of the project schedule and budget.

A constructability review and value analysis should be performed as part of each internal design review as well. Constructability addresses the efficiency with which the system can be installed. Value analysis refers to determining if the owner's needs and requirements can be met using alternative materials, equipment, and systems at a lower cost. Constructability reviews and value analyses are most effective when conducted early in the design process.

On green design-build projects, each internal design review should also include a review of the project's green goals to ensure that these goals are being met by the design. Where the owner's project criteria requires that the building perform to a certain level or achieve a certain certification, the green design review may require that simulations be run with the updated design to ensure that it still meets the requirements. Also, if a point system is being used to rate the building's sustainability, the criteria for awarding points or credits should be reviewed to ensure that the necessary number of points or credits would be earned by the building to meet the owner's project criteria.

Owner Design Reviews. Owner design reviews should be scheduled in accordance with the agreement between the owner and contractor. Owner design reviews are essential because the contractor keeps the owner informed of the design and its progress through these reviews. Owner design reviews should be scheduled only after corresponding internal design reviews are completed and should involve representatives throughout the owner's organization that are affected by the project.

5.13 MANAGING THE DESIGN PROCESS

5.13.1 Establishing Design Documentation Requirements

The detail in the design documentation will vary from project to project. The contractor needs to determine the level of detail needed to obtain permits and efficiently construct the building. However, as the level of design detail increases, so does the design cost. Given this trade-off between installation efficiency and design cost, the contractor needs to determine the level of design detail that is appropriate for the project.

For a simple project, the design could stop as early as the schematic design stage and be adequate. However, for larger, more complex projects, a detailed design with a full set of drawings and specifications will probably be required. The contractor needs to decide the level of detail based on the trade-off between time spent by installation crews designing the work as it is installed and the cost of the more-detailed design. There is no right or wrong answer as to how much design documentation is needed.

5.13.2 Design Change and Modification Procedure

The contractor should establish and document design change and modification procedures with the owner at the start of the project. These procedures are important and could be a part of the contractor's quality assurance program for projects if it has one. The owner's business is dynamic, and before the design is completed, changes in owner requirements may occur that necessitate changes to the system design. These changes must be documented and agreed upon by the owner's designated representative, because they might impact not only the technical performance of the system but also the project schedule and budget. In addition, the contractor should be compensated for significant design changes.

5.13.3 Plan and Schedule the Design Process

Planning and scheduling the design process on a design-build project is as important as planning and scheduling the system installation. The contractor needs to pay as much attention to the design process as it does to construction, because on a design-build project, design delays can delay project completion the same as construction delays. Planning the design process involves scheduling the completion of design stages as well as milestones, such as when information needs to be provided by the owner and vendors and when internal and owner design reviews need to take place, among other things.

5.14 CASE STUDY

SpawGlass

SpawGlass Office Building, Texas

After lengthy discussions with Kirksey on the benefits and costs associated with a green building, leaders of SpawGlass, a 100 percent employee-owned general contractor, decided to "go green."

The former SpawGlass office bulding was sold quickly once it was on the market. The buyers of the former building let SpawGlass know that they would need to move in seven months later. That left only seven months for the entire design and construction process of the new building. The SpawGlass construction team broke ground on the new building in July 2002, and the firm, which is ranked among the Best Companies to Work for in Texas, moved in six months later in January 2003.

The office building, the first LEED-certified building in Houston, is 20,000 square feet. The building can accommodate 75 employees, in both enclosed offices and cubicles. The SpawGlass building has a white roof for reflectivity, uses extensive natural light to reduce energy costs, and low-E glass to reduce heat gain, glare, and energy loss. Exposed concrete floors are located in the lobby, kitchen, and corridors.

Other elements include concrete grade beams, slab on grade, tilt-up precast concrete walls, and structural steel framing. Glazed aluminum curtain walls at the main entry and north elevation provide natural light. Glazed aluminum clerestory windows at each of the floor bays provide additional lighting.

Kirksey adapted conventional design methods to create an energy-efficient design. Despite the challenging seven-month schedule from design to completed construction, both SpawGlass and Kirksey committed to this endeavor—a pioneer facility in Houston's green movement and a milestone in both companies' histories.

Figure 5-1 Photo Courtesy of Aker/Zvonkovic Photography.

Working in a building with natural light pouring in is just one of the many benefits of a green building. The building design and construction has created a healthy, productive environment for SpawGlass.

The long sides of the building face north and south, to avoid direct sunlight. Exposed overhead duct work means less sheet rock and dust particles. SpawGlass recycled 75 percent of construction site materials, including concrete, metal, paper, and wood. This project served as a catalyst for instituting a recycling program at other SpawGlass jobsites and at the main office. Even the bathrooms were designed to save on energy and water, including waterless urinals.

Figure 5-2 Photo Courtesy of Aker/Zvonkovic Photography.

The following are LEED criteria and just a few ways in which SpawGlass achieved them:

Indoor Environmental Quality

Ductwork was protected during the construction process to minimize particulate accumulation within the system.

Low-VOC (volatile organic compounds) products were used, including paint, adhesives, carpet, and sealants.

Windows are located throughout the building to provide daylight and views for more than 75 percent of occupied spaces.

Water Efficiency

Native plants and a weather-monitoring device were installed to help reduce demand for landscape irrigation.

Efficient plumbing fixtures were selected to reduce the calculated building water usage by 37 percent over a conventional building.

Water-saving fixtures include motion-sensor lavatory faucets, waterless urinals, and low-flow kitchen faucets and shower heads.

Energy and Atmosphere

An independent building commissioning agent confirmed electrical and HVAC performance requirements, preventing any inefficiencies during and after installation.

The building is calculated to be 56 percent more efficient than the baseline standard as a result of building orientation, window protection, a light-colored roof, and high-efficiency electrical and HVAC systems.

Materials and Resources

A construction waste and recycling plan salvaged an estimated 75 percent of the construction debris and created a profit for the project.

Products that were manufactured or fabricated locally include: Glass, gypsum board, structural steel, roof insulation, metal fascia, ductwork, concrete, millwork, paint, ceramic tile, iron and PVC pipe and fittings, concrete masonry, canopies, and landscape mulch.

Sustainable Sites

An erosion control plan was developed and implemented to protect the stormwater system.

Vegetated swales and bio-retention basins were designed to reduce the rate of stormwater runoff and remove water contaminants.

An Energy Star-compliant roof system and paving surfaces with a high reflectance were specified to minimize heat absorbed on the site.

Figure 5-3 Photo Courtesy of Aker/Zvonkovic Photography.

In striving for LEED certification, the budget had to be continuously monitored to keep costs in perspective. Many cost options associated with becoming green were analyzed and budgeted before making a decision to proceed. The team calculated anticipated energy cost savings on various systems and materials. Payback analyses were performed on several design issues, and decisions were made based on the practicality of cost.

With a fast-track schedule, material deliveries were monitored daily in order to achieve the schedule. A one-day turnaround on shop drawings and daily architect site visits were instrumental in achieving the six-month construction time frame. As a design-build project, the design was completed as the job was being built. Changes were made continually throughout the process.

Lessons learned

1) Utilizing design-build on our project was critical for its success. Not only was the project extremely fast track, but we were also aiming to become the first LEED-certified building in Houston. Like most design-build projects, we started construction before the final completion of the design. Unique to our building, however, is that we were also making some of the "green" decisions on the fly. We had to work closely with the designer at Kirksey to decide which LEED points to pursue. The designer would propose an idea for a LEED point and we would quickly evaluate it and, if it made economical sense, we would implement it. The project would probably have taken twice as long under a traditional system.

2) Construction waste management turned out easier than expected and actually reduced the cost of waste disposal. At each stage of construction, we only had to set out two to three recycling dumpsters to cover the predominant types of waste during that stage. Once the workers understood the program, they were very good about putting the waste in the correct dumpster. Our cost for construction waste disposal was significantly reduced because we did not have to pay for landfill fees for waste that was sent to recyclers. Not only that, but we were also reimbursed for all the metal waste that we sent. On this relatively small project, we figured that we saved over $4,000 in the costs of construction waste.

5.15 REFERENCES

Associated General Contractors of America (AGC), *Standard Form of Design-Build Agreement and General Conditions between Owner and Design-Builder*, AGC Document No. 410, 1999. [Refer also to pending

ConsensusDOCS: General Contracting Number 410 entitled Owner/Design-Builder Agreement & General Conditions (Cost Plus With A Guaranteed Maximum Price), 2007.]

Associated General Contractors of America (AGC), *The Contractors' Guide to BIM*, First Edition.

Design-Build Institute of America (DBIA), Design-Build Definitions, *Design-Build Manual of Practice*, Document Number 103, October 1996.

Design-Build Institute of America (DBIA), *Standard Form of General Conditions of Contract between Owner and Design-Builder*, Document Number 535, 1998.

chapter 6

Green Subcontracting

6.1 INTRODUCTION

Subcontracting portions of the project work to specialty contractors on green construction projects results in some unique challenges for the contractor. The success of a green construction project depends on subcontractor performance, which means that specialty contractors must understand their role and responsibilities. This chapter focuses on the unique challenges faced by the contractor that is subcontracting work on a green project. This chapter covers subcontractor qualifications and prequalifying subcontractors for green projects, defining the subcontract scope of work on green projects, educating subcontractors about their responsibilities, green subcontract terms and conditions, and training the subcontractor's craftworkers during construction.

6.2 GREEN SUBCONTRACTING

A subcontractor is usually a specialty contractor that has unique knowledge, skills, and resources and that can perform a portion of the work at the project site more efficiently and effectively than the contractor. AGC Document No. 200 entitled *Standard Form of Agreement and General Conditions between Owner and Contractor* defines a *subcontractor* in subparagraph 2.3.16 as follows:

> A Subcontractor is a person or entity retained by the Contractor as an independent contractor to provide the labor, materials, equipment and/or other services necessary to complete a specific portion of the Work.

Therefore, the contractor enters into a contract with the specialty contractor to perform a portion of the contractor's scope of work. This contract is referred to as a subcontract, and the specialty contractor is referred to as a subcontractor. Unlike a material or equipment supplier, a subcontractor actually performs work at the project site under the direction of the contractor. Managing procurement and material and equipment suppliers is addressed in Chapter 7.

The subcontractor might be a firm that manufactures or fabricates materials, assemblies, or equipment off-site for incorporation into the project at the site and also performs the installation as part of its subcontract with the contractor. An example of this is the sheet metal contractor that fabricates ductwork in its shop, ships it to the project site, and then installs it with its own forces. Prefabrication is becoming more common in the construction industry as other trades prefabricate materials and assemblies off-site and then install them on-site. Even though fabrication is performed off-site, the contract that the contractor has with the specialty contractor is a subcontract, and the specialty contractor is a subcontractor.

Similarly, a subcontractor could be a firm whose primary business is the fabrication or manufacture of materials and equipment and is also responsible for the installation of the material or equipment at the site in its subcontract with the contractor. In this case, the subcontractor could contract with another specialty contractor to actually perform the installation at the project site. An example of this would be a steel fabricator that subcontracts the steel erection portion of its contract to a steel erector. Similarly, this would also be the case where the manufacturer of the building management system subcontracts the installation of the raceway system, cable installation, and associated branch circuiting to an electrical contractor. In both cases, the firm that has a contract directly with the contractor is referred to a subcontractor or first-tier subcontractor. The firm that has a contract with the subcontractor to perform a portion of its work is referred to a subsubcontractor or second-tier subcontractor. AGC Document No. 200 defines a *subsubcontractor* in subparagraph 2.3.18 as follows:

> A Subsubcontractor is a person or entity who has an agreement with a Subcontractor to perform any portion of the Subcontractor's Work.

Figure 6-1 illustrates the contract chain through second-tier subcontractors and suppliers.

Because most materials and equipment procured for the project will be procured by subcontractors and most of the work at the project site will be performed by subcontractors, inexperienced or unqualified subcontractors can put the contractor at risk. The contractor's responsibility for the performance

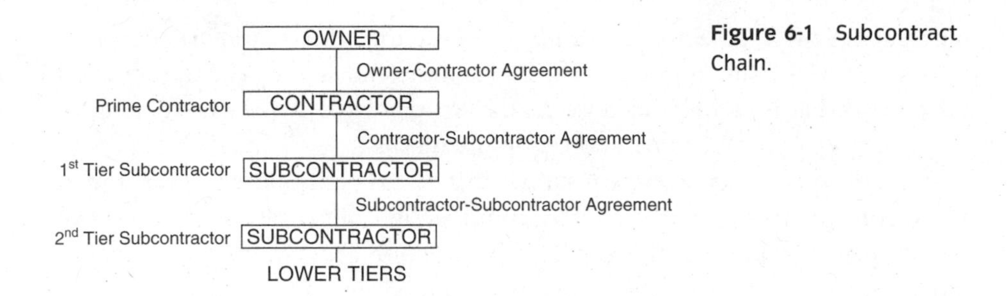

Figure 6-1 Subcontract Chain.

of its subcontractors is usually addressed in the construction agreement between the owner and contractor. For example, the *Standard Form of Agreement Between Owner and Contractor*, which is AGC Document No. 200, in Subparagraph 3.4.2 requires the following:

> The Contractor shall be responsible to the Owner for acts or omissions of parties or entities performing portions of the Work for or on behalf for the Contractor or any of its Subcontractors.

Therefore, the contractor is responsible not only for the performance of its first-tier subcontractors but also for the performance of subsubcontractors or second-tier subcontractors, along with any other specialty contractors that subcontract with second-tier subcontractors, and so on.

6.3 WHY ARE SUBCONTRACTORS IMPORTANT ON GREEN PROJECTS?

Subcontractors perform mostof the construction work on building projects, and the contractor is responsible for their performance as discussed previously. The size, function, and complexity of a commercial or institutional building project will determine the number of subcontractors on a project; this can range from as few as a dozen subcontractors to hundreds. The ability to effectively manage subcontractors on a traditional building project will often determine the success of the project not only for the contractor but also for everyone involved, including the owner and subcontractors.

On green building projects, the ability of the contractor to achieve the project's sustainable objectives depends on the subcontractors' performance and the contractor's ability to effectively manage subcontractors. The contractor must be able to communicate the green project objectives and requirements to the subcontractors as well as educate them about their role in achieving these objectives. During construction, the contractor needs to make sure that the subcontractors' field personnel and craftworkers are trained and know what they need to do on a day-to-day basis at the site to achieve the project's green objectives and requirements.

The subcontractors need to know that a green building is as much about construction practices as it is about design, and they have a crucial role to play throughout the project. In addition, subcontractors need to understand that sustainable construction is an ongoing process that starts with preconstruction and finishes at closeout. With green construction projects, the risk of subcontractor performance is magnified, because the success of a green project is heavily dependent on subcontractors buying in to the process and actively working toward achieving the project's green objectives and requirements.

6.4 SPECIALTY CONTRACTOR PREQUALIFICATION

Specialty contractors should be prequalified for green building projects whenever possible. Instances when it is not possible to prequalify specialty contractors might be on public building projects or where the owner has included designated subcontractors in the project specifications. In these cases, it is important that the contractor assesses the subcontractor's knowledge and understanding of green building construction and develops a strategy for dealing with specialty contractors that may not be qualified. Prequalification is particularly important when subcontractors are selected based on price through a competitive bidding process.

The purpose of subcontractor prequalification is to screen out unqualified specialty contractors that do not have the background and experience to effectively participate as a member of the contractor's construction team. AGC Document No. 621 entitled *Subcontractor's Statement of Qualifications* provides a standard format for gathering information about potential subcontractors. AGC Document No. 621 is broken down into 11 sections that can be modified to address potential subcontractor experience on green construction projects. Figure 6-2 provides an outline of AGC Document No. 621. Included with AGC Document No. 621 are the following three schedules:

Schedule A: Key Construction Personnel

Schedule B: Past Projects

Schedule C: Current Construction Projects

Section I	Subcontractor's Organization
Section II	Licensing & Registration
Section III	Subcontractor's Personnel & Management Approach
Section IV	Subcontractor's Relevant Experience
Section V	Subcontractor's Relevant Experience
Section VI	Surety & Insurance
Section VII	Subcontractor Financial Information
Section VIII	Industry Agreements, Affiliations, Memberships, Awards, & Honors
Section IX	Statement Of Potential Conflicts Of Interest
Section X	Other Information
Section XI	References
Schedule A	Key Construction Personnel
Schedule B	Past Projects
Schedule C	Current Construction Projects

Figure 6-2 AGC Document No. 621 Contents.
Subcontractor's Statement of Qualifications for a Specific Project.

The contractor can modify AGC Document No. 621 for green building projects or use it as a guideline to develop its own form for determining the potential subcontractor qualifications for green building projects. For example, Section III addresses the subcontractor's personnel and could be expanded to ask if the specialty contractor has personnel who are certified by outside third parties. An example of this would be subcontractor personnel who have been certified by the U.S. Green Building Council as Leadership in Energy and Environmental Design Accredited Professionals (LEED-AP). In addition, the contractor could ask specifically about the amount of green building experience that the specialty contractor's personnel have.

The specialty contractor is also asked in Section III what criteria it will use to select subsubcontractors on this project. In addition to the standard criteria listed, such as price and financial strength, the subcontractor could be asked if it will make experience on green construction projects a criterion for subsubcontractor selection. Furthermore, the specialty contractor is also asked in this section about its proposed technical and managerial approach to the construction project; this could be expanded to ask about its approach to sustainable construction and meeting green contract requirements, such as third-party certification as a green building.

Section IV asks about the subcontractor's relevant experience on past and current projects. In addition to the standard questions about annual construction volume and project performance, this section could also ask specifically about green construction projects that the specialty contractor has completed in the past. Similarly, Section VIII asks about industry affiliations and memberships as well as industry awards and honors. This section could be modified to specifically ask about memberships and active participation in local and national organizations that are involved with or promote green construction as well as any awards and recognition for green construction or energy-efficient systems that the subcontractor has received. A cover letter could be sent with the standard AGC Document No. 621 that would modify it to solicit information from potential subcontractors about their experience and capabilities in green building construction.

6.5 DEFINING SUBCONTRACTOR SCOPE

Clearly defining the subcontract scope is critical to the success of any construction project for the contractor. Subcontractor scope must not only be clearly delineated but its definition must also ensure that there are no gaps or overlaps. On green building projects, scope definition can be even more challenging, because important requirements that need to be included in the subcontractor's scope of work may not be included or referenced in the CSI *MasterFormat*™ specification division or section that is normally assigned to the subcontractor.

The purpose of the Construction Specifications Institute (CSI) *MasterFormat*™ is to improve communication on construction projects by providing a clear and consistent method for organizing project manuals and specifications. The CSI *MasterFormat*™ has been used on most commercial and institutional building projects since its introduction in 1963. The CSI *MasterFormat*™ is commonly used by contractors to break down a project and define subcontractor scope, even though both the 1995 and 2004 editions that are currently in use state the following (MasterFormat™ 1995 and 2004):

> MasterFormat's organizational structure used in a project manual does not imply how the work is assigned to various design disciplines, trades, or subcontractors. MasterFormat is not intended to determine which particular elements of the project manual are prepared by a particular discipline. Similarly, it is not intended to determine what particular work required by the project manual is the responsibility of a particular trade. A particular discipline or trade is likely to be responsible for subjects from multiple Divisions, as well as from multiple Subgroups.

Not only is the CSI *MasterFormat*™ not intended to provide a breakdown for defining subcontractor scope, but there is also no requirement that the specifier use the list of numbers and titles that make up the CSI *MasterFormat*™. This is especially true for green building requirements that are general and not specifically tied to a particular product or work associated with a particular specialty contractor's scope of work such as construction waste management. Many green construction requirements apply to multiple CSI *MasterFormat*™ divisions and sections and are either specified in Division 01, which covers general requirements, or by reference to other specifications sections not normally included in a particular subcontractor's scope of work. In defining subcontractor scope, the contractor must do a thorough review of the project plans and specifications to ensure that all green requirements have been identified and included in the appropriate subcontractor's scope of work.

As is discussed in the subcontract terms and conditions section of this chapter, the subcontractor may be contractually bound to meet green requirements via the flow-through clause or by references to other contract documents in the subcontract. The contractor should call attention to these requirements in its request for bid or proposal (RFP) to subcontractors to ensure that the subcontractor knows about these requirements and has incorporated them into its bid or proposal. Not having these requirements covered in the bid or proposal may result in disputes during construction, which can strain relationships and mar an otherwise successful green project.

Experienced subcontractors can assist the contractor in scope definition, especially when the contractor is operating as a construction manager or design-builder and is able to bring key specialty contractors into the project at an early stage. The contractor should utilize specialty contractors whenever possible to help analyze a green project and define subcontractor scopes of work.

6.6 BASIS FOR GREEN SCOPE OF WORK

When defining the subcontractor scope of work for green building projects, the starting point should be the traditional project breakdown. This breakdown varies across the country as well as from contractor to contractor. This traditional breakdown may not pick up all of the green requirements or even most of the green requirements, depending on how the construction documents are structured and how green requirements are incorporated into those documents.

After completing the initial scope of work breakdown, the next step is to look for green building requirements in other specification sections that may add to the subcontractor's scope of work. This review is very much like the review that is performed on traditional projects to identify problems in the contract documents so they can be resolved, as well as gaps and overlaps between subcontractors.

If the project is seeking third-party certification as a green building based on a set of criteria or if there are other explicit green building requirements, such as construction criteria and building performance requirements, these requirements must be included in the subcontractor's scope of work when applicable. These requirements must be reviewed to determine their impact on the subcontractor's scope of work and included if these criteria will impact the subcontractor's scope of work. For example, if the project is required to be certified by an outside third party to a particular level, the specific requirements should be reviewed to determine the impact on each subcontractor's scope of work.

6.7 ESTABLISHING GREEN CRITERIA

If the contract documents are vague in defining green criteria for the project, then the contractor needs to define the criteria. For example, the specifications may require that the project be certified to a certain level using a third-party green building rating system, but the criteria for achieving that certification may not be specified. Because the specific criteria will impact the subcontractor's scope of work and subcontractor pricing, specific requirements that will be met need to be identified by the contractor in advance and included in the appropriate subcontractor's scope of work.

6.8 GREEN PROCESSES AND PROCEDURES

Based on the green requirements incorporated into the project, the contractor needs to define various processes and procedures to ensure that the green requirements are met. These processes and procedures will often impact the subcontractor's scope of work and will need to be incorporated into the subcontractor's scope of work for bidding. An example of a green procedure would be how the contractor plans to handle and account for construction waste on the project in order to achieve goals for recycling and credit from third-party green building rating systems for water management. Because subcontractors typically generate a significant amount of recyclable waste on a construction project, it is imperative that they understand the project goals associated with construction waste recycling, the procedure for recycling waste at the project site, how to account for the amount of waste recycled, and how to provide this information to the contractor.

6.9 EDUCATE SPECIALTY CONTRACTORS

The contractor cannot assume that subcontractors understand their roles and responsibilities on a green building project. Even if the subcontractors have previous experience with other green building projects, they may not be familiar with the goals of the current project that may impact their work, the contractor's processes and procedures for achieving these goals, or the documentation required to be submitted that verifies what the subcontractor has done to help the contractor achieve the project goal. Furthermore, even if the subcontractor understands the project goal regarding the green building requirement that the contractor is trying to achieve and follows the contractor's process for achieving that goal but fails to document its efforts properly, the project may not get credit for the subcontractor's efforts, and it may impact the project's green building rating.

The contractor must educate subcontractors about their roles and responsibilities on a green building project before bidding in order to make sure that they know what is expected of them and that any additional work associated with these green building requirements is included in their bids. Among other things, subcontractors should be educated regarding the project's green requirements associated with the following:

- Contract administration
- Procurement
- Material delivery and storage
- Construction processes and procedures
- Start up and commissioning
- Material, equipment, and system documentation
- Subcontract closeout

6.10 SUBCONTRACTOR SELECTION

Once subcontractors are qualified, the process used to select subcontractors on a green building project should be no different than that for a traditional building project. The first step in the selection process would be to determine the criteria for selecting subcontractors for the project. On a green building project, this selection criteria may be different for different subcontractors depending on their role and responsibilities associated with meeting and documenting the project's green building criteria. For example, the HVAC subcontractor typically has a much bigger role in meeting green building criteria than a drywall subcontractor. Therefore, the contractor may elect to select the HVAC subcontractor based on a combination of green building qualifications and price, with price being a secondary issue because of the key role that the HVAC subcontractor plays in a green building project. The contractor's past experience working with the HVAC subcontractor as well as the HVAC subcontractor's ability to work effectively with others, such as the owner, mechanical engineer, controls contractor, and commissioning authority, should be an important consideration. Because the drywall work on a green building project is usually not as critical as the HVAC system installation and commissioning, the drywall subcontractor could be selected from a group of qualified contractors based solely on price.

Once the selection criteria have been determined, the contractor needs to request subcontractor bids or proposals based on the selection criteria. The selection criteria need to be communicated to the specialty contractors, particularly if factors other than price will be considered. If the selection criteria are not adequately communicated to specialty contractors, then their proposal may not adequately address the contractor's criteria, and the contractor may not select the best specialty contractor for the project because it does not have a complete picture of the subcontractor's qualifications and ability to help the contractor be successful on this project.

Once proposals have been received, the contractor then needs to evaluate the specialty contractor proposals against the selection criteria. The evaluation of specialty contractor proposals should consider not only qualifications and price but also that the specialty contractor understands its role and responsibilities for achieving the project's sustainable goals and third-party rating system documentation requirements. Based on this evaluation, the contractor should select the preferred specialty contractor for the project.

During negotiation and before executing a subcontract with the preferred specialty contractor, the contractor should review the project's green goals, the specialty contractor's role and responsibilities, the green construction processes and procedures, and any required testing or documentation. This is important so that no misunderstanding occurs between the contractor and subcontractors regarding the green project requirements and their impact on the subcontractor's scope of work and cost of performing the work.

6.11 SUBCONTRACT TERMS AND CONDITIONS

6.11.1 Current Model Contract Documents

Currently, the model subcontracts do not specifically address green building requirements. This section discusses some standard subcontract terms and conditions that address green requirements either directly or indirectly. This section is based on AGC Document No. 450 entitled *Standard Form of Agreement between Contractor and Subcontractor*, but the provisions of this model subcontract are common to most subcontracts.

6.11.2 Subcontract Documents

The subcontract documents include applicable portions of the contractor's contract documents plus any other documents specified as part of the subcontract and mutually agreed upon by the contractor and subcontractor. These documents are typically listed in the subcontract as illustrated by paragraph 2.3 of AGC Document No. 650:

> **2.3 SUBCONTRACT DOCUMENTS** The Subcontract Documents include this Agreement, the Owner-Contractor agreement, special conditions, general conditions, specifications, drawings, addenda, Subcontract Change Orders, amendments and any pending and exercised alternates. . . . The Subcontract Documents existing at the time of the execution this Agreement are set forth in Article 13.

6.11.3 Green Flow-Through Requirements

On any construction project, a chain of contracts ties all parties from the owner down to the lowest-tier subcontractor, as illustrated in Figure 6-1. In order to ensure consistency in each contract tier, flow-through clauses are incorporated into the contract that links each successive tier to those above it. These clauses bind parties on one contract tier to the applicable contract requirements of the next-higher tier. Flow-through clauses can be very important on green construction projects because, as noted previously, there is no set place where green requirements appear in the owner-contractor agreement, which includes not only general, supplemental, and special conditions of the contract, but also the drawings and specifications as well as addenda before contract award and change orders after contract award.

Paragraph 3.1 of AGC Document No. 605 provides an example flow-through clause:

> **3.1 OBLIGATIONS** The Contractor and Subcontractor are hereby mutually bound by the terms of this Subcontract. To the extent the terms of the prime contract between the Owner and Contractor apply to the work of the Subcontractor, then the Contractor hereby assumes toward the Subcontractor all the obligations, rights, duties, and redress that the Owner under the prime contract assumes toward the Contractor. In an identical way, the Subcontractor hereby assumes toward the Contractor all the same obligations, rights, duties, and redress that the Contractor assumes toward the Owner and Architect under the prime contract. In the event of an inconsistency among the documents, the specific terms of this Subcontract shall govern.

Paragraph 3.2 further outlines the subcontractor's responsibilities to the contractor and requires the subcontractor to furnish its best skill and judgment in the performance of the Subcontract Work and to cooperate with the Contractor so that the Contractor may fulfill its obligations to the Owner. Helping the contractor fulfill its obligations to the owner includes adhering to the contractor's green processes and procedures at the project site, which cover such things as construction waste management.

In addition, each tier of contracts usually requires that the contractor or subcontractor tie its lower-tier subcontractors to the applicable requirements to which it is bound as well. For the contractor, an example of this requirement appears in Paragraph 5.3 of AGC Document No. 200:

> **5.3 BINDING OF SUBCONTRACTORS AND MATERIAL SUPPLIERS** The Contractor agrees to bind every Subcontractor and Material Supplier (and require every Subcontractor to so bind its subcontractors and material suppliers) to all the provisions of this Agreement and the Contract Documents as they apply to the Subcontractor's and Material Supplier's portions of the Work.

6.11.4 Implicit Green Requirements

Up until now, green building requirements have been assumed to be explicitly required by the contract documents, which means they are specifically called out in a particular specifications section even if it is Division 01 that covers general requirements. However, sometimes green requirements are not explicitly called out in the contract documents but are explicitly required by reference to a third-party green building rating system, certification or labeling of materials or equipment, specific codes and standards, or local, state, or federal laws. For example, Division 01 might require compliance with a specific code such as the International Code Council's (ICC) 2006 *International Energy Conservation Code*, or simply requiring compliance with all applicable codes and requirements in the jurisdiction where the work is to be performed.

Federal, state, and local governments are increasingly enacting laws and statutes that require buildings with certain characteristics to be green buildings. These laws and statutes often use third-party rating systems as a basis for determining green requirements. Subcontracts typically include language that requires the subcontractor to comply with all federal, state, and local laws in the performance of its work; this includes the new green building statutes and laws. An example of this is Paragraph 3.29 of AGC Document No. 650 as follows:

> **3.2.9 COMPLIANCE WITH LAWS** The Subcontractor agrees to be bound by, and at its own costs comply with, all federal, state and local laws, ordinances and regulations (the Laws) applicable to the Subcontract Work, including but not limited to, equal employment opportunity, minority business enterprise, women's business enterprise, disadvantaged business enterprise, safety and all other Laws with which the Contractor must comply. The Subcontractor shall be liable to the Contractor and the Owner for all loss, cost and expense attributable to any acts of commission or omission by the Subcontractor, its employees and agents resulting from the failure to comply with Laws, including, but not limited to, any fines, penalties or corrective measures, except as provided in Subparagraph 3.14.9.

6.11.5 Subcontractor Green Submittals

In addition to the usual shop drawings, samples, product data, manufacturers' literature, and other submittals typically required, the subcontractor will need to make other submittals on a green building project. These submittals will be determined by the green building criteria or rating system being used on the project and the scope of work of the subcontractor. Examples of green building submittals include the following:

- List of Salvaged and Refurbished Materials
- List of Recycled Materials

- List of Regional Materials
- List of Certified Wood Products

In addition to submitting lists of materials, the contractor also needs to make sure that subcontractors know when the project requirements require them to submit material cost reports and receipts to show the percentage that a particular type of material, such as a regional material, is used on the job to meet green requirements.

6.11.6 Subcontractor Design Management Responsibilities

Chapter 5 covered design management for the contractor on green design-build projects. Green building projects may be design-build or involve a performance specification for a particular building system that requires the subcontractor to perform design work. If the subcontractor needs to retain an outside designer, then the subcontractor's responsibilities associated with providing the design must be included in the subcontract. Paragraph 3.8 of AGC Document No. 650 addresses subcontractor design responsibilities and requirements:

> **3.8.1** If the Subcontract Documents (1) specifically require the Subcontractor to provide design services and (2) specify all design and performance criteria, the Subcontractor shall provide those design services necessary to satisfactorily complete Subcontract Work. Design services provided by the Subcontractor shall be procured from licensed design professionals retained by the Subcontractor as permitted by the law of the place where the Project is located (the Designer). The Designer's signature and seal shall appear on all drawings, calculations, specifications, certifications, Shop Drawings and other submittals prepared by the Designer. Shop Drawings and other submittals related to the Subcontract Work designed or certified by the Designer, if prepared by others, shall bear the Subcontractor's and the Designer's written approvals when submitted to the Contractor. The Contractor shall be entitled to rely upon the adequacy, accuracy and completeness of the services, certifications or approvals performed by the Designer.
>
> **3.8.2** If the Designer is an independent professional, the design services shall be procured pursuant to a separate agreement between the Subcontractor and the Designer. The Subcontractor-Designer agreement shall not provide for any limitation of liability, except to the extent that consequential damages are waived pursuant to Paragraph 5.4, or

> exclusion from participation in the multiparty proceedings requirement of Paragraph 11.4. The Designer(s) is (are) ____. The Subcontractor shall notify the Contractor in writing if it intends to change the Designer. The Subcontractor shall be responsible for conformance of its design with the information given and the design concept expressed in the Subcontract Documents. The Subcontractor shall not be responsible for the adequacy of the performance or design criteria required by the Subcontract Documents.

6.11.7 System Startup and Commissioning

Most subcontracts require the subcontractor to assist the owner and contractor in the startup and commissioning of the building systems for which the subcontractor is responsible. Paragraph 3.28 of AGC Document No. 650 illustrates a typical subcontract requirement for assisting with system startup and commissioning:

> **3.28 SYSTEM AND EQUIPMENT STARTUP** With the assistance of the Owner's maintenance personnel and the Contractor, the Subcontractor shall direct the check-out and operation of systems and equipment for readiness, and assist in their initial startup and the testing of the Subcontract Work.

Because system startup and commissioning as well as project closeout activities are more extensive on green building projects (as discussed in Chapter 9), the contractor will want to review the subcontractor's contract documents to make sure they reflect the actual requirements of a green building project rather than those of a traditional building project. Where necessary, the system and equipment startup clause of the subcontract should be modified to address the additional startup and commissioning requirements for a particular project.

6.11.8 Early Startup of Building Systems

Early startup of building systems can impact warranties. For example, the warranty period for HVAC equipment, such as chillers, boilers, pumps, fans, and other similar equipment, usually starts when the equipment is started up. In a traditional building, this is not a problem because the HVAC systems are usually not started up until substantial completion is near, and the manufacturer's warranty normally corresponds with the contractor's one-year contractual construction warranty on installed systems and equipment. This is illustrated by paragraph 3.21 of AGC Document No. 650:

> **3.21 WARRANTIES** The Subcontractor warrants that all materials and equipment furnished under this Agreement shall be new, unless otherwise specified, of good quality, in conformance with the Subcontract Documents, and free from defective workmanship and materials. Warranties shall commence on the date of Substantial Completion of the Work or a designated portion.

However, with green buildings, HVAC equipment and systems may need to be started earlier for commissioning and startup requirements, as well as for conditioning interior spaces and indoor air quality during construction. This may require not only purchasing an extended warranty on the HVAC equipment that is started up early but also cleaning, changing filters, and other work that is not normally required before turning over the green building to the owner. Subcontractors need to be aware of these early startup and other green building requirements that may impact their warranties.

6.12 TRAINING SUBCONTRACTORS ON-SITE

As noted earlier in this chapter, it is very important that subcontractors be educated regarding their role and responsibilities on the green building project. Equally important, each subcontractor's workforce must also understand what green construction is all about and their role in achieving the project's green objectives. Problems can arise for the contractor when the subcontractor's craftworkers don't understand their role and responsibilities on a green project. The contractor must make sure that the subcontractor's craftworkers are trained and know not only what they are supposed to do but also what they are not supposed to do.

For instance, the craftworkers need to understand the contractor's recycling goals, what materials are recyclable, where to put materials that are recyclable, and how to document that materials have been recycled. The best construction waste management plan is worthless without the buy-in and implementation of the subcontractor's craftworkers. Similarly, the crews need to understand that adhesives, sealants, paints, coatings, and other everyday consumables used on a green construction project may adhere to specific volatile organic compound (VOC) requirements to meet project air-quality standards both before and after occupancy. If a consumable such as a single tube of sealant that does not meet the low-VOC requirements is used, the entire project could lose credit for the use of low-VOC materials. This simple mistake could adversely impact the level of third-party green building certification achieved by the project, or the work may need to be redone by the subcontractor with an acceptable low-VOC product.

The contractor may want to require that all subcontractors submit a plan for training craftworkers and monitoring their crews to ensure that field personnel are adhering to the contract requirements and the contractor's green processes and procedures. This training could take place in conjunction with or in a similar manner to toolbox safety talks. If the subcontractor is new to green construction, the contractor may want to assist the subcontractor in preparing a training plan, provide training materials, and even supply personnel to conduct the training. These training sessions should be conducted regularly throughout the construction process to maintain awareness. In addition, field personnel and crews change during the course of a project as craftworkers come and go with the workload, so repeating training on important topics should also be considered.

6.13 CASE STUDY

The Beck Group

The Shangri La Botanical Gardens and Nature Center, Texas

The Shangri La Botanical Gardens and Nature Center will be a world-class, Leadership in Energy and Environmental Design (LEED™) platinum-level facility that will put Orange, Texas, on the map. The project is funded by the Nelda C. and H. J. Lutcher Stark Foundation, whose mission is to improve the quality of life in Southeast Texas. Shangri La is nestled within the Stark Foundation's 262-acre property in the heart of the City of Orange. The nature center was designed by Lake Flato Architects, Inc., Mesa Design Group, and Jeffrey Carbo, ASLA. The Austin office of the Dallas-based Beck Group was awarded preconstruction and construction services for the $15 million project. The Historical Garden is a reflection of the Stark Family's Botanical Garden which closed in the 1950s after a fierce winter storm decimated the landscape. Ironically, the 18-month construction schedule to bring the property back to its original beauty was scrapped in 2005, in the wake of the devastation caused by Hurricane Rita during the early stages of construction.

Figure 6-3 Nature Discovery Center.
Photo courtesy of The Beck Group.

Figure 6-4 Photo courtesy of The Beck Group.

Green Attributes

The landscape design team had planned the new gardens around a heavy tree canopy. The project had been designed to specifically utilize the shade to minimize the heat island effect. This was required for one of the LEED™ credits that were calculated into the project's potential rating. However, the once heavily vegetated property lost 70 percent of its tree canopy when Hurricane Rita came through. Cleanup was a long and arduous process.

The Shangri La site reclamation efforts were ongoing and focused on tactical deadfall extraction and preservation of surviving trees. Mobile milling equipment from Austin was staged in a centralized location on the Shangri La property. The owner agreed to salvage the deadfall timber for a feature in the historic garden called the Cypress Gate. The structure is made of four 30-foot-tall cypress trees and other timber milled onsite. Log benches and furniture for the visitor center, boardwalks, and outdoor classrooms were also milled from deadfall caused by Hurricane Rita.

The visitor center incorporates five individual buildings that are designated as environmentally responsible, profitable and healthy places to visit and work. The structure's concrete foundations consist of 40 percent fly ash. The walls and ceilings are insulated with soy-based insulation. The majority of the walls are built of reclaimed brick material salvaged from a warehouse in Arkansas that was built in 1910. The roofing is designed to reflect heat and collect rainwater runoff for use in toilets and in the extensive landscape irrigation system. There are also photovoltaic panels installed on a portion of the south-facing roofs that are contributing to renewable energy credits.

The mechanical equipment is connected to a closed-loop, geothermal heating and cooling system, which pumps water from an 800-foot-deep well, benefiting from the earth's reduced underground temperature. Well water, or thermal transfer fluid, is distributed through underground piping to each of the building's geothermal/water source heat pumps, used for either cooling or heating. After flowing through the system, the water is returned underground. The buildings use low-flow fixtures and water-conserving technologies that will contribute to two water use reduction credits. The project is on track to divert over 80 percent of all construction debris from the landfill, and we anticipate receiving all of the Low-Emitting Material credits. The project also used a high percentage of recycled content materials and locally manufactured and extracted materials.

Figure 6-5 Photo courtesy of The Beck Group.

Lessons Learned

One of the most important criteria for delivering sustainable construction and all of the documentation that it entails is early and ongoing training and support of the subcontracted trades. It is still widely believed that green building has cost premiums associated with it, so it is imperative to help overcome this perception by creating an easily understood documentation methodology that is shared with the subcontractors during the bidding phase of a project that is then followed by a preconstruction conference after buy-out that revisits all of the sustainability requirements and the procedures to be used.

The Beck Group has several LEED™-rated projects that have allowed us to create knowledge transfer and evolving best practices to assist the next awarded green or LEED™ project. While sharing team members between projects is a rare luxury, the sharing of information and internal training and lessons learned does not have to be. Beck has created a documentation system with forms that is shared with every subcontractor so that they do not have to try to figure out how and when to meet a project's sustainability needs. An example of the best-case scenario for generating subcontractor investment and involvement in the green building process is cost and product information shared to the best of their ability during preconstruction with actual cost and proper submittal documentation per LEED™ provided shortly after buy-out. If this sequence is not encouraged or enforced, the time required to revisit not only each subcontractor, but also each product or system, will cause costly delays and uncertainties regarding the project's ability to meet specified goals.

A Unique Texas Refuge

When complete, the nature center will offer a unique environment where visitors can learn about and explore the distinctive ecosystems within the eastern-most part of Texas. Arriving at any of the three educational outposts, visitors can experience the diverse flora and fauna within the cypress/tupelo swamp and see more than 300 species of plants. Completion is scheduled for the late fall of 2007. To learn more, or to visit this unusual ecological haven, go to www.shangrilagardens.org.

6.14 REFERENCES

The Associated General Contractors of America, Standard Form of Agreement and General Conditions Between Owner and Contractor (Where the Contract Price Is a Lump Sum), AGC Document No. 200, 2000. [Refer also to pending ConsensusDOCS: General Contracting Number 200 entitled Owner/Contractor Agreement & General Conditions (Lump Sum), 2007.]

The Associated General Contractors of America (AGC), AGC Document No. 621, Subcontractor's Statement of Qualifications for a Specific Project, 2004. [Refer also to pending ConsensusDOCS: General Contracting Number 721 entitled Subcontractor's Statement of Qualifications, Specific Project, 2007.]

The Associated General Contractors of America (AGC), AGC Document No. 650, Standard Form of Agreement between Contractor and Subcontractor, 1998. [Refer also to pending ConsensusDOCS: General Contracting Number 750 entitled Contractor/Subcontractor Agreement (Contractor at Risk), 2007.]

Construction Specifications Institute, *MasterFormat*™, Alexandria, Virginia, 1996.

Construction Specifications Institute, *MasterFormat™ 2004 Edition*, Alexandria, Virginia, 2004.

Sheet Metal and Air Conditioning Contractors' National Association, Inc. (SMACNA), SMACNA Position Paper, *Early Start-Up of Permanently Installed HVAC Systems*, Undated.

chapter 7

Green Procurement

7.1 INTRODUCTION

Materials and permanently installed equipment are critical in green building construction and represent a major portion of criteria used to classify or certify a green building. Even though the design team specifies the materials and equipment that will be incorporated into the building, the contractor and its subcontractors must understand the material and equipment specifications as well as the characteristics that make the specified materials and equipment green. This makes material and equipment procurement a critical success factor in any green construction project. This chapter addresses procuring material and equipment for green building project.

7.2 BUILDING PRODUCT LIFE CYCLE

Figure 7-1 illustrates the life cycle of a building product from raw material extraction or harvesting to reuse, recycling, or disposal. Understanding the building product life cycle is very important to understanding the procurement and use of materials and equipment for green building projects. The following sections detail each of the steps in the building product life cycle.

7.2.1 Raw Material Extraction/Harvesting

The life cycle of a building product starts with the extraction or harvesting of the raw material that will be used to produce the finished building product. Products such as stone or wood are primarily single-material building products, although they may use other materials in their manufacture. The term *extraction* refers to removing a raw material from the earth, such as stone, that cannot be replenished. *Harvesting* refers to the acquisition of raw materials, such as wood, that are usually plant-based materials and can be replenished over time.

From a green building standpoint, this distinction is very important. The use of raw materials that need to be extracted from the earth can result in environmental impact during the extraction process, including impacting the earth itself, affecting water supplies both underground and above ground, and possibly polluting the atmosphere. In addition, the use of these virgin materials deprives future generations of their use. Whether the contract documents call for this level of attention or not, the use of extracted materials should be kept to a minimum in green building construction.

Harvested materials can be regrown and replenished, although the harvesting of plants, such as trees, for wood products can also impact the environment. The earth's ecological balance depends on plant life, and the

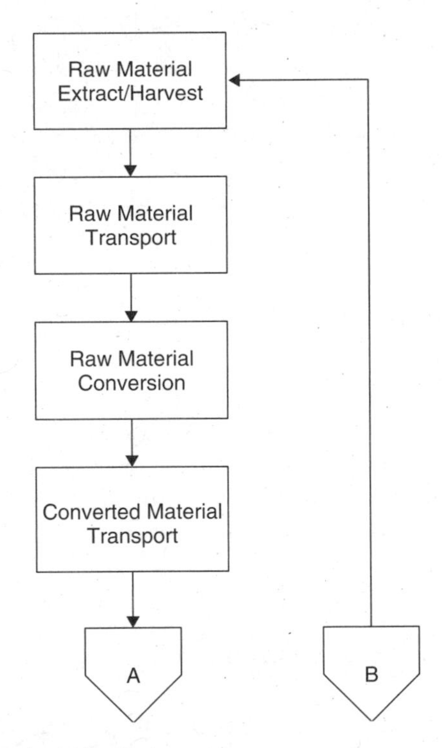

Figure 7-1 Building Product Life Cycle.

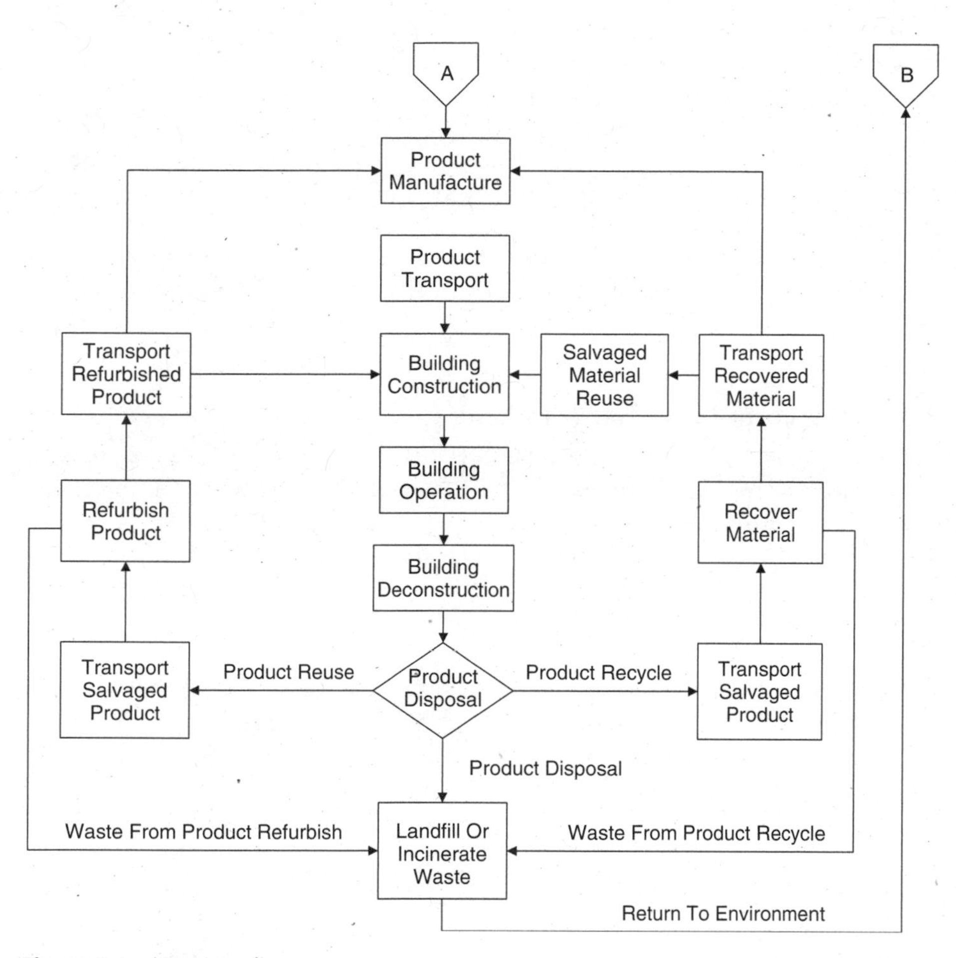

Figure 7-1 *(Continued)*

harvesting of trees and other plants can impact air quality as well as animal life and soil erosion. Furthermore, different types of plants require different periods of time to reach maturity. For example, bamboo can be substituted for wood because it will replenish very quickly after harvesting, whereas hardwoods such as oak can take decades to mature to the point where they are ready to be reharvested.

7.2.2 Raw Material Transport

Raw material transport refers to the need to transport the raw material from the location where the raw material is extracted or harvested to where it will be converted from a raw material into a bulk material that can be used to manufacture a final building product.

7.2.3 Raw Material Conversion

This step in the building material life cycle converts the raw material or materials into a bulk material that can be used to manufacture a building product.

7.2.4 Converted Material Transport

Following raw material conversion, the converted bulk material is transported to the location where the final building product is manufactured.

7.2.5 Product Manufacture

In this step, the final building product is manufactured. Like other steps in the building product life cycle, this step may include multiple manufacturing processes and the transport of the product in process between various manufacturing locations.

7.2.6 Product Transport

The manufactured building product is transported from the manufacturing facility to the building construction site. Again, this step may include several intermediate stops at manufacturer distribution centers, construction material distributor facilities, and even retail building product outlets. The final product transport includes all transportation and storage between the location where the final product is manufactured and the building construction site.

7.2.7 Building Product Installation

Once the building product is delivered to the project site, it can be incorporated into the building. This step is triggered by custody of the material changing hands and becoming the responsibility of the installing contractor. This step not only includes the actual incorporation of the building product into the building project but also storage on or off-site by the contractor, as well as movement between interim storage sites and to the location where the product will actually be installed.

7.2.8 Building Product Use

Building operation represents the use of the building product as a functioning part of the building over the building's useful life or the useful life of the building product, whichever is shorter.

7.2.9 Building Product Removal

Building product removal can occur during or after the building's useful life. During a building's life, deconstruction can include the removal and replacement of one or more building products. The removal and replacement of one building product might occur as a result of (1) the product's useful life being less than that of the overall building, as in the case of a boiler; (2) technological obsolescence, where it is more economical to replace the existing building product with a more efficient product, as in the case of windows; or (3) simply a change in occupancy or style, as in the case of interior finishes, such as wall or floor coverings.

The renovation of all or part of a building is an example of when deconstruction would require the systematic removal of more than one building product. Similarly, at the end of its useful life, the entire building would be deconstructed, which involves the systematic dismantling and removal of building products for either recycling, reuse, or disposal.

Both building deconstruction and demolition achieve the same end: the removal of either part of or all of an existing building that has reached the end of its useful life. However, building deconstruction differs from demolition in that deconstruction is the systematic dismantling of the building with the objective of reusing or recycling as much of the building materials as possible. Building demolition's primary objective is to remove the existing building from the site to make room for new development. While building material recycling or reuse also happen with building demolition, they are secondary to the main objective of removing all or part of the building.

7.2.10 Building Product Disposition

As part of building deconstruction, a decision needs to be made as to what should be done with the building product or products that are removed from the existing building. A removed building product can be reused, recycled, or disposed of. The disposition of a particular building product will first depend on owner-contactor agreement and whether it places any specific requirements on the disposition or ownership of materials removed by the contractor during demolition. If not and the contractor owns the removed building products, then the contractor needs to determine the most economical way of disposing of these materials. For materials that can be economically reused, the contractor may consider the sale of these materials and use the proceeds of the sale to offset some of its deconstruction costs making it more competitive. For materials that will either be recycled or disposed of, the contractor needs to determine the most economical method of removing these materials from the project site.

7.2.11 Transport Salvaged Product

Whether the product is to be reused or recycled, it will need to be transported from the building deconstruction site to the location where it will either be refurbished if it is going to be reused or material recovered if it is going to be recycled. The transportation from the building deconstruction site to the location where the building product will be refurbished or recycled might be across town or across the globe. This transportation cost plays an important consideration as to whether it is economically feasible to reuse or recycle a building material.

7.2.12 Refurbish Product

In this step, the building product is refurbished in order to be reused for either its intended purpose or for another purpose. An example of refurbishing a product and reusing it for its intended purpose might be lighting fixtures. Antique gas or electric lighting fixtures are often rewired and outfitted with lamps that retain the fixture's character and provide an interesting architectural feature in a new or renovated space. Similarly, relatively new lighting fixtures that are being removed from an existing building that have good photometrics for an application can be cleaned and inspected and then rewired and relamped as required with energy-efficient lamps and ballasts and reused in a new application. A building product can also be refurbished and used for another purpose. For example, casework could be removed from an existing building that is either being razed or renovated for another purpose and used in a new building for a similar or different purpose. Similarly, bricks and roof tiles from buildings being razed can be refurbished and reused.

Waste from refurbishing a building product that cannot either be reused or recycled would be sent to a landfill or incinerated. Any portion of the waste that results from product refurbishing that could be recycled and reused should be recovered and reused.

7.2.13 Transport Refurbished Product

The refurbished building product can either be transported directly to a building construction site for installation or enter the building material distribution process and be stocked by a building material distributor until needed. Once purchased, the building product would be taken from stock by the distributor, transported to the project site, and incorporated into the building construction.

7.2.14 Recover Material

Once the building product is delivered, the material recovery process begins. This might be a one-step process for a homogeneous building product where the useful material is recovered directly. Alternately, it might be a multistep process where the building product is first broken down into its various recyclable components, and then each of these components is processed at the same facility or transported to other facilities specializing in a particular material for recycling.

7.2.15 Transport Recovered Material

Once the bulk material is recovered, it is transported from the recovery facility to the manufacturing facility, where it is incorporated into a new building product.

7.2.16 Landfill/Incinerate Waste

Building products removed from the building being deconstructed as well as waste from the product refurbishing or material recovery processes are returned to the environment by either landfilling or incinerating the waste. The goal is to minimize this waste, because landfilling and incinerating waste typically have a negative environmental impact.

7.3 WHERE ARE THE GREEN BUILDING PRODUCT REQUIREMENTS?

7.3.1 Specific Requirements

A project's green building product requirements should be clearly defined in the contract documents. Ideally, the green building product requirements will be incorporated into their associated drawings and specification sections. However, green building product requirements are not always explicitly called out in the drawings and specifications. The contractor needs to know what the green building product requirements are for the project and the impact they will have on the contractor's procurement and construction processes. It is very important that the contractor communicate these explicit or implicit requirements to subcontractors, because subcontractors may not be aware of the unique requirements of a green building product (as discussed in Chapter 6).

7.3.2 General Performance Requirements

If the green building product requirements are not explicitly included in the drawings and specifications, then where can the contractor expect to find them? Green building product requirements can be incorporated into the contract documents in a variety of ways. For example, the green project requirements could be included in Division 01 and simply require that the building be certified or certifiable to a particular level using a given third-party green building rating system. For example, the bid documents might require that the project be certified to the silver level using the LEED™ *Green Building Rating System for New Construction and Major Renovations* (LEED-NC). This would mean that the building would need to earn at least 33 out of 69 possible points plus meet the requirements of category prerequisites. This would essentially amount to a performance specification (as discussed in Chapter 2), and the contractor would then determine how to achieve the required number of points for certification, which would most likely impact its procurement process.

7.3.3 Partial Performance Requirements

The owner could specify the criteria and associated points that it has already met through design and the additional criteria that it expects to be met during construction to achieve the desired level of certification. Alternatively, the owner could leave part or all of the criteria needed to achieve the desired level of certification up to the prime contractor. In either case, the contractor needs to understand its responsibility for providing green building products.

7.3.4 Need to Understand Green Building Product Requirements

The contractor needs to understand green building material requirements and its responsibilities for meeting those requirements. This includes not only the project managers and estimators but also the contractor's workforce and subcontractors. During bidding, it is very important that estimators know and understand green building product requirements. During construction, everyone must understand the project's green building product requirements and their responsibilities for meeting those requirements. For example, if a sealant or adhesive is used indoors after building enclosure that does not meet the requirements for using low-emitting materials, then the project could lose credit for this requirement or the contractor could be required to replace the material, resulting in additional rework that impacts both the project schedule and cost.

7.4 SUBCONTRACTOR PROCUREMENT RESPONSIBILITY

Subcontractors do most of the procurement on a green building project, and they are key to the success of a green building project (as discussed in Chapter 6). Therefore, it is very important that subcontractors understand that material and equipment procurement is critical to achieving the goals and objectives of a green building project. Furthermore, subcontractors need to know that expendables such as sealants, adhesives, and paint that may not be required to be submitted for review and approval must meet green building requirements whether called out in the specifications or not. As noted in the previous section, using an expendable that does not meet the green building requirements could bar the owner from receiving credit toward certification or require removal and rework by the subcontractor. In addition, subcontractors must understand their responsibilities regarding record keeping and documentation of the materials and equipment that they install.

7.5 GREEN BUILDING PRODUCT CHARACTERISTICS

General green building product characteristics include the following:

- Resource efficiency
- Waste minimization
- Indoor air quality
- Energy efficiency
- Water conservation

The following sections address each of these general green building product characteristics. The definitions of each of these green building characteristics, as well as the criteria for including products in these categories, can vary greatly based on who is defining them and the purpose, geographic location, and third-party green building rating system, among others. The category definitions and criteria are not absolute, and the contractor needs to be aware that the definition and criteria used on one project may be different on another project.

The purpose of this section is to provide the contractor with a framework for understanding green product categories that are often referred to in project specifications and third-party green building rating systems and not to provide definitive definitions or criteria. On any project, the contractor needs to look to the contract documents and the third-party green building rating system being used for the specific green product definitions and criteria that apply to that particular project.

7.5.1 Resource Efficiency

Resource Efficiency Defined. Resource efficiency is about producing a building product with a minimum of resources and waste. Figure 7-2 provides a simple input-output model for the building product production process. In this simple model, raw materials and energy are used to produce and deliver the

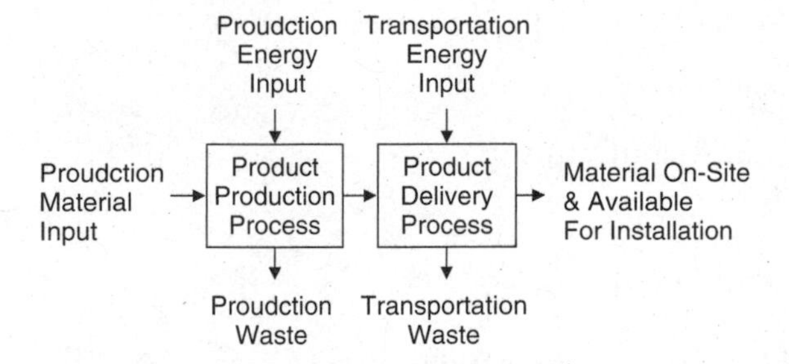

Figure 7-2 Building Product Production Process.Input-Output Model.

finished building product to the building site for incorporation into the work. It should be noted that this simple two-step building product supply model comprising production process and delivery process actually represents the entire building product supply chain, which would contain a myriad of sub-processes and activities depending on the building product being addressed.

Limitations of Current Focus on Resource Efficiency. Figure 7-2 is focused on resource efficiency or reducing the material and energy inputs to the building product production process through production process design and waste minimization. A broader discussion of resource efficiency in the production of building products would necessarily address the use of other resources, such as labor, production equipment, and capital, that are used to produce a green building product. This is especially the case where increased efficiency in terms of raw material and energy used in the production of a green building product is achieved through the substitution or increased use of other resources. However, today's focus on the use of green building products in the construction industry, including third-party green building rating systems, does not explicitly account for factors other than material and energy efficiency. Material and energy use in the production of green building products is easier to both understand and measure than society's alternate use of other resources. These other factors that include labor, production equipment, and capital are left to the free-market pricing system, which includes the impact of government policy on the price of a green building product when compared to a standard building product.

Resource Efficiency Metrics. As can be seen from Figure 7-2, resource efficiency in green building products addresses input materials, input energy, building product production, and building product delivery. Building product resource efficiency can be measured using the following eight metrics:

- Product Recycled Content
- Product Salvaged, Refurbished, or Remanufactured Content

- Product Natural, Plentiful, or Renewable Resource Content
- Product Local or Regional Content
- Product Resource-Efficient Manufacturing
- Product Durability
- Product Reuse or Recycle Potential

The following sections address each of these eight metrics.

Product Recycled Content. Product recycled content is measured as either postconsumer or postindustrial content by third-party green building rating systems. Materials with postconsumer recycled content are made from raw materials that are composed partially or totally of household waste, such as soda cans, milk jugs, newspapers, and other waste. Materials with postindustrial recycled content are made from raw materials that are composed partially or totally of industrial waste, such as synthetic gypsum, fly ash, and blast furnace slag. Most third-party green building rating systems encourage the use of products with identifiable recycled content, including postindustrial content with a preference for postconsumer content. Use recycled materials to reduce the use of raw materials and divert materials from landfills.

Product Salvaged, Refurbished, or Remanufactured Content. Building products that fall into this category have been salvaged, refurbished, or remanufactured before being incorporated into new building construction. Designers often find these products very difficult to incorporate into the specifications because of their uniqueness and the need to obtain them in advance. These materials are often prepurchased by the owner and supplied to the contractor for installation. An example of a product that would fit in this category would be ornate doors that are taken out of a building being demolished and reused in a building undergoing major renovation.

Product Natural, Plentiful, or Renewable Resource Content. These materials include products such as certified wood. Certified wood is harvested from sustainably managed sources and is usually certified as such by an independent third party using guidelines such as the Forest Stewardship Council's (FSC) Principles and Criteria Guidelines (FSC 2007) or The Sustainable Forestry Board (SFB) and American Forest & Paper Association's (AF&PA), Sustainable Forestry Initiative Standard (SFIS): 2005-2009 Standard (SFB 2004). Other rapidly renewable materials that can be grown and harvested for raw materials in a specified cycle can be used to reduce the depletion of virgin materials and the use of petroleum-based materials. The contractor may be required to include rapidly renewable materials and certified wood in the project.

Product Local or Regional Content. Products in this category include building materials, components, and systems found locally or regionally within a given radius of the project site. The objective of using these local and regional building materials is to reduce transportation to the project site and support the local and regional economies.

Product Resource-Efficient Production. These products are manufactured in a resource-efficient manner, which includes reduced energy consumption, minimized waste by using recycled or recyclable materials, and reduced product packaging, among other things.

Product Local or Regional Production. Similar to the use of products with local and regional content, this category of green building products addresses the use of locally manufactured materials and equipment. Typically, these building products are defined as being produced within a given radius of the building site, and the goal is to reduce transportation costs and support the local or regional economy.

Product Durability. This category refers to the selection of a product that uses a material that is more durable than the typical product that would normally be used. As a result, the more durable product will not have to be replaced as often over the life of the building as the typical product saving materials and energy.

Product Reuse or Recycle Potential. Last, but not least, is the selection of products that can be easily reused or recycled into new building products at the end of their useful life rather than just disposed of.

7.5.2 Waste Minimization

Waste minimization strategies that can be implemented by the contractor during the building product procurement process include the following:

- Order only what is actually needed.
- Minimize shipping and packing materials.
- Use standard-size materials wherever possible.
- Consider custom-fabricated materials.
- Consider prefabricating material assemblies off-site.

The following sections discuss each of these waste minimization strategies.

Order Only What Is Actually Needed. To reduce waste, order only what is needed. This will usually require preplanning, close coordination with suppliers, and ongoing monitoring of upcoming work, production rates, and on-site inventories. Ordering only what is needed is essentially just-in-time supply chain management, and it will reduce waste, theft, and spoilage as well as the contractor's investment in inventory and inventory carrying costs. The risk of ordering only what is necessary is material shortages, which can result in lost time and productivity as well as the cost of idle labor and equipment.

On-site vendor-managed inventory (VMI) is one way the contractor can ensure that crews have the materials they need when they need them, and there is no overordering that eventually becomes waste. VMI requires that the contractor partner with its supplier by negotiating a project supply agreement for materials and the supplier setting one or more trailers on-site to house the materials. The supplier keeps the trailers stocked with the materials currently needed on the project and invoices the contractor monthly for materials used. At the end of the project, the supplier removes the trailers with the remaining materials, and the contractor does not have any excess materials it needs to store or dispose of or pay a supplier restocking fee. In addition, the supplier can be responsible for disposing of all of the shipping and packing materials.

Another example of this strategy would be if a project calls for wood framing. The framing component of the project can be ordered after a material take-off and cut list has been generated. Having the material delivered from a cut list saves materials from being wasted on the project site and saves time during the execution of the work in the field.

Minimize Shipping and Packing Materials. Wherever possible, the contractor should work with suppliers to minimize product shipping and packing materials. The best strategy is to specify reusable or returnable product shipping and packing materials whenever possible. This will eliminate waste and the need to dispose of shipping and packing materials on-site.

Use Standard-Size Materials Wherever Possible. The contractor should review the project plans to determine how compatible the project design is with standard-size building products. Wherever possible, the contractor should try to use standard-size materials to reduce the need for field fabrication and adjustments that lead to waste. The contractor can lay out the project using standard-size materials and request changes from the design team to make standard-size building products fit where possible.

Consider Custom-Fabricated Materials. If standard-size materials will not work or are not acceptable, the contractor can work with its suppliers to get materials with the needed dimensions even though they may be nonstandard. Custom-fabricated materials may cost more, but they will reduce waste and improve field productivity.

Consider Prefabricating Material Assemblies Off-Site. Off-site fabrication is becoming more common and can reduce waste at the jobsite as well as increase productivity. Off-site fabrication of ductwork, piping, and other assemblies can be performed in a controlled environment and then the prefabricated assemblies delivered and installed at the site. Not only does this reduce waste, but off-site prefabrication in a controlled environment can also reduce the risk of contamination at the site. In addition, off-site fabrication in a controlled environment may allow the use of paints, sealants, or adhesives that could not be used on-site because of construction air-quality concerns.

7.5.3 Indoor Air Quality

Indoor air quality (IAQ) can be improved during and after construction by using green building products and strategies with the following characteristics:

- Low-toxic or nontoxic
- Minimal chemical emissions
- Moisture-resistant
- Capable of being healthfully maintained

The following sections address each of these building product characteristics that contribute to a healthy indoor environment for both construction workers and building occupants.

Low-Toxic or Nontoxic. These materials or consumables emit few or no carcinogens, reproductive toxicants, or irritants as demonstrated by the manufacturer through appropriate testing.

Minimal Chemical Emissions. Products that meet the minimal chemical emissions criteria are those that have minimal volatile organic compound (VOC) emissions. The project specifications or the third-party green building rating system will establish limits on chemical emissions that products will need to meet. Usually, these limits are established by reference to specific standards, such as the following:

- South Coast Air Quality Management District Rule No. 1168 for adhesive VOC limits (SCAQMD 2007)

- Bay Area Air Quality Management District Regulation 8 Rule No. 5 for sealant VOC limits (BAAQMD 2007)
- Green Seal requirements that provide VOC and chemical component limits for paints and coatings (Green Seal 2007)
- The Carpet and Rug Institute's Green Label and Green Label Plus testing and labeling for low-VOC carpet, cushion, and adhesives (CRI 2007)

Moisture-Resistant. Products in this category resist moisture in order to inhibit the growth of biological contaminants in buildings.

Healthfully Maintained. These products require only simple, nontoxic, or low-VOC methods of cleaning.

7.5.4 Energy Efficiency

Products, including building appliances and equipment, that fall into this category are designed to be energy efficient and are often required to be tested and certified to meet a specified standard. An example of an energy-efficiency program is Energy Star®, which is jointly administered by the U.S. Environmental Protection Agency and the U.S. Department of Energy. The purpose of Energy Star is to protect the environment through energy-efficient products and practices (EPA and DOE 2007).

7.5.5 Water Conservation

Building products that fall into this category conserve or eliminate the use of water in their operation. As with energy-efficient products, these products have been tested and meet a specified standard.

7.6 GREEN BUILDING PRODUCT SCREENING PROCESS

In a straightforward product specification, the green building product screening process should be completed by the designer and reflected in the material specifications. However, when products have not been adequately screened by the designer, the contractor may have to perform the screening process. The green building screening process involves the following three steps, as illustrated in Figure 7-3 [Froeschle 1999]:

- Step 1: Research
- Step 2: Evaluate
- Step 3: Select

The following sections discuss each of these three steps.

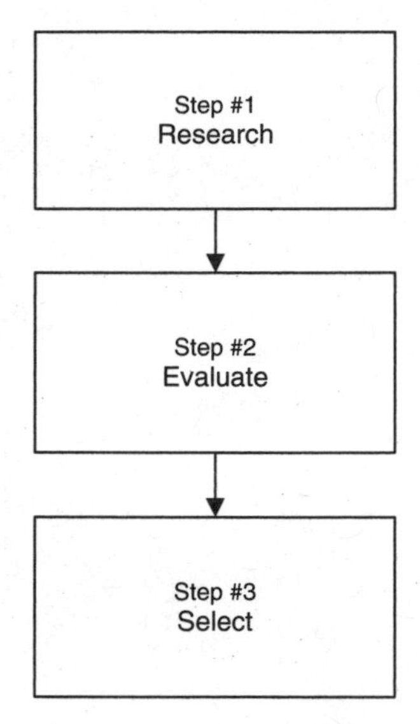

Figure 7-3 Green Building Product Screening Process.

7.6.1 Step 1: Research

This step involves gathering technical information on the building product to be evaluated. This involves information such as the Material Safety Data Sheets (MSDSs), IAQ test data, product warranties, source material characteristics, recycled content data, environmental statements, and durability information. In addition, this step should include the researching of building codes and regulations to determine if the material being researched can be used.

7.6.2 Step 2: Evaluate

Evaluation involves the review and confirmation of the technical information to ensure that it is complete and meets the project requirements. Evaluation of multiple building products is straightforward when comparing building products based on the same environmental criteria (e.g., a comparison of recycled content among various products). Evaluation becomes more complex when comparing different building products that perform the same function. The outcome of this step will be a set of acceptable building products that will be compared in Step 3.

7.6.3 Step 3: Select

The final step in the selection process involves comparing the characteristics of a set of acceptable building products identified in Step 2 and selecting the preferred building product for installation.

7.7 GREEN PROCUREMENT PROCESS

In general, the procurement of materials and equipment is very similar to the procurement process on a traditional building project. However, the contractor needs to be aware of and address some additional steps and requirements in its procurement process. The green procurement process can be summarized as the following three subprocesses:

- Green building product requirements analysis
- Supplier RFQ development
- Product procurement

Each of these subprocesses is linked and is covered in the following sections of this chapter.

7.7.1 Green Building Product Requirements Analysis

As discussed previously, green project requirements can be found just about anywhere in the contract documents. This is also true for green building product requirements that are an important part of any green building project. Just like green project requirements, green building product requirements can be included in the contract documents either explicitly or implicitly.

Green building product requirements are expressed explicitly when the required green product characteristics are included in the product's respective specification section with other standard product requirements. Implicit green building requirements are usually included in the contract documents by reference. Also, explicitly including green building product requirements should speed the bid process, increase bid accuracy, provide a level playing field for both experienced and inexperienced bidders, and result in fewer change orders and disputes during construction and commissioning.

Explicitly including green building requirements in the specification and on the drawings where appropriate is the preferred approach, because this lowers the contractor's risk. In the broadest sense, green building product requirements could be included directly in the owner-contractor agreement, as part of the supplemental or special conditions of contract, or in Division 01, which is the general specifications section whose requirements apply to all other specific specification sections unless excluded. These general green requirements could simply state that the project is to be certified or certifiable to a certain level using a specific third-party green building rating system. Then it is up to the contractor to determine how this requirement is going to be met and what role green building products will play in achieving the green building objectives. Recently, some federal agencies and state and local governments have begun requiring that public and private buildings under their jurisdiction meet predetermined green building requirements, which is another way that green building product requirements can be incorporated implicitly into the contract documents.

All this means that the contractor must be vigilant regarding green building product requirements and their impact on the procurement process. The contractor should not assume that green product requirements will be found exclusively in each building product's respective specification section along with traditional technical requirements. Nor should the contractor assume that the products that are called out in the specifications will have the same weight when evaluating them for their green characteristics. Instead, green building product requirements may be included elsewhere in other parts of the contract documents. This increases the contractor's risk, and the contractor must carefully analyze the contract documents to identify the green building product requirements for which it will be specifically responsible. A process for analyzing green building product requirements is illustrated in Figure 7-4. The following sections describe this process.

Review and Identify Green Building Product Requirements. As can be seen from Figure 7-4, the first step in the process is to review the contract documents to determine where the green requirements can be found. Once the green requirements are found, they can be categorized as one of the following:

- General requirements
- Specific requirements
- Mixed requirements

Furthermore, from the previous discussion, it can be seen that the general requirements correspond to implicit green project requirements and specific green requirements correspond to explicit green project requirements. Green project requirements can also be mixed, with some green building products being specified explicitly and others being specified implicitly.

General Green Requirements Path. The first step in the general green product requirements path is to identify specific green building product requirements. In other words, what green building products need to be procured in order to meet the project's overall green goals and objectives. In the case where the contract documents simply state that the building will be a green project, the contractor will need to review the green building criteria incorporated into the contract documents to determine how that criteria will impact procurement and what building products need to be green.

Once specific building products are identified as needing to be green, the next step is for the contractor to establish specific product requirements

Figure 7-4
Green Building Product Requirements Analysis.

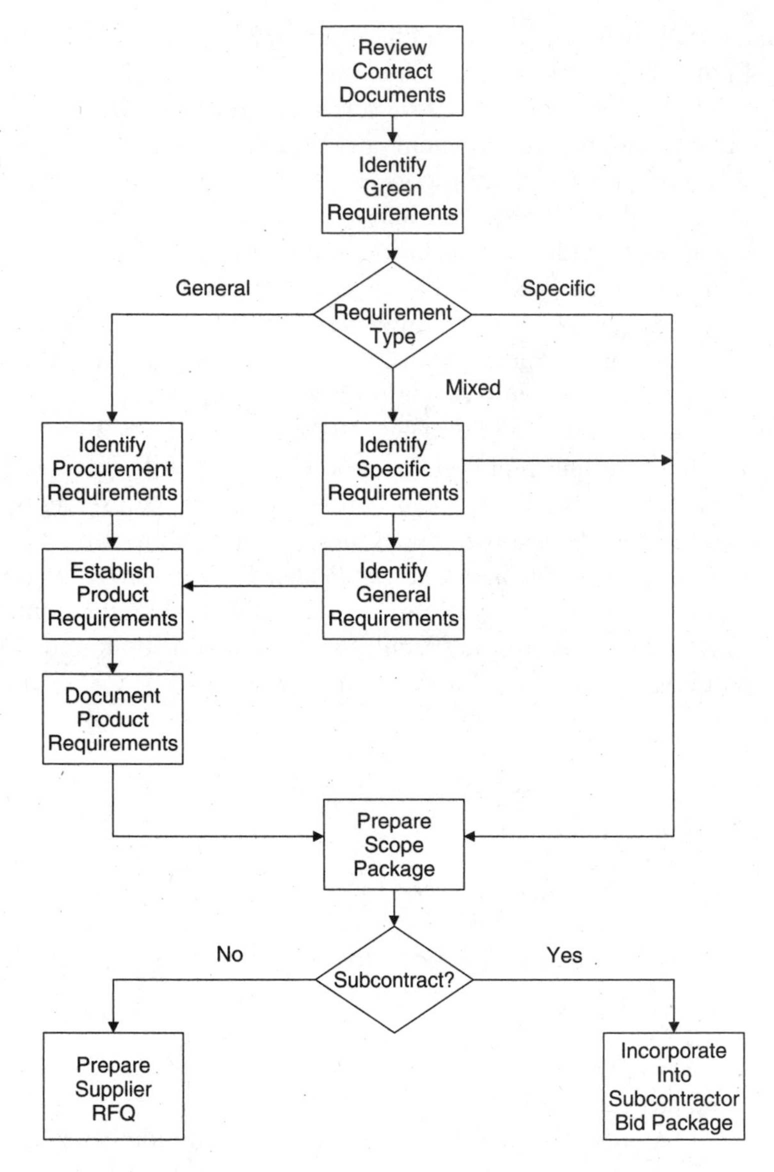

for these building products. The specific building product requirements will include any of the green building product characteristics discussed earlier in this chapter. The nature of the building product and the green building criteria that the building product must meet will determine which of the green building product characteristics need to be met. The specific green building product requirements identified need to be documented so they can be either procured by the contractor or incorporated into the request for subcontractor quotation if a specialty contractor will be used to perform the work.

Mixed Green Requirements Path. Mixed requirements include green building product requirements that are both general and specific. Most green building projects fall into this category, and the contractor needs to separate the general and specific requirements to ensure that they are addressed properly. As noted previously, the mixed green requirements path requires that the contractor determine whether green building product requirements are either general or specific and then enter the appropriate path. Green building products that are specified explicitly using performance, descriptive, or prescriptive specifications can be used directly to prepare project bid packages. Green building requirements or products that are specified in general need to enter the general green requirements path, and specific green building product requirements need to be determined and documented before these requirements can be used to prepare a bid package.

Specific Green Requirements Path. With the specific green building requirements path, the bid documents already contain specific green building product requirements as discussed previously, and these specific requirements can flow directly into the preparation of the scope of work that will be self-performed by the contractor or subcontracted to a specialty contractor.

Prepare and Distribute Bid Package. Once the green building product requirements have been identified and documented, they can be incorporated into bid packages for the project work. At this point, the contractor can decide what project work it will self-perform and what project work it will subcontract to specialty contractors. The green building product requirements that are included in the bid packages that the contractor will self-perform can now be used to develop supplier requests for quotation (RFQs). Similarly, those bid packages that will be subcontracted can be sent to specialty contractors, who in turn can use the green building product information to develop RFQs for green building products to send to their suppliers. It is critical to include all requirements as part of the RFQ, including both general and specific green requirements.

7.7.2 Supplier RFQ Development

The supplier RFQ process for green building products is illustrated in Figure 7-5. This process starts where the green building product requirements analysis shown in Figure 7-4 leaves off. At the completion of the green

Figure 7-5 Supplier RFQ Development.

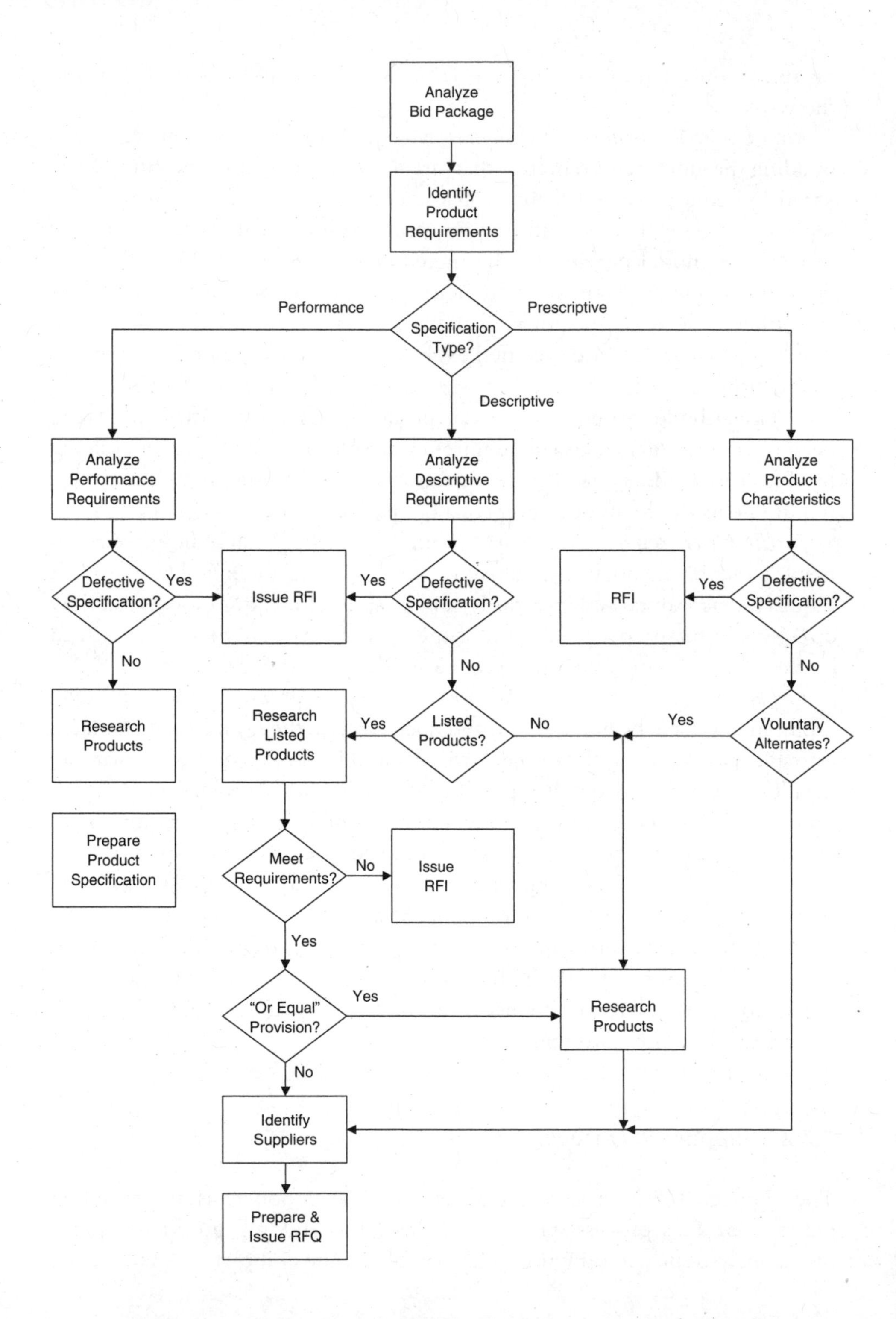

building product requirements analysis, the contractor has divided the project into several bid packages that encompass the overall project scope and decided which of these bid packages it will self-perform and which it will subcontract to specialty contractors. Each of these bid packages has a clearly defined scope of work, which includes procuring the building products necessary to perform the work. The supplier RFQ process covered in this section is applicable to either the contractor or subcontractor that needs a supplier quotation for a building product that will be installed as part of its scope of work and part of its bid or proposal.

The supplier RFQ development process is very important, because the supplier quotation received as a result of the RFQ becomes a part of the contractor's project bid or proposal, whether the contractor or a subcontractor obtains it. In addition, the RFQ is the basis for not only the supplier's price quotation but also the quoted product's performance characteristics, which are key for a green building project. The following sections discuss the steps in the supplier RFQ development process.

Identify Product Requirements. The first step in the supplier RFQ development process is to analyze the bid package. For the contractor, this step may have already been completed and skipped for the work it plans to self-perform. However, if the team that divided the project up and developed bid packages to be used as the basis for the contractor's bid or proposal is not the same as the person or group in the contractor's organization that will obtain quotations from suppliers for needed building products, then this step may need to be performed by the contractor's procurement personnel.

If the bid package will be subcontracted to a specialty contractor to perform, then the specialty contractor will need to analyze the bid package and identify building product requirements in order to effectively request supplier quotations. Once building product requirements have been identified, the next step is to determine if these product requirements are expressed as performance, descriptive, or prescriptive specifications. As can be seen from Figure 7-5, the type of specification will determine the sequence of activities that need to be performed to prepare an RFQ for qualified suppliers. The following sections describe the step-by-step RFQ procedure involved in each of these paths.

Performance Specification Path. If the green building product requirements are expressed as a performance specification, then the next step in the supplier RFQ process is to thoroughly analyze the performance specification. This analysis should determine not only how the product is to perform but also what performance risks the contractor or subcontractor might be assuming. For example, does the performance specification include unusually stringent performance requirements, new and untested technology, design of systems or subsystems, unconventional performance guarantees, extended warranties, or other nonstandard technical requirements? If there are any questions concerning the validity of the performance specification or clarification concerning the performance specification requirements, the contractor should issue a request for information (RFI) to clarify the building product requirements.

The contractor should also make sure that green building products specified using a performance specification include measurable performance criteria. It is best if this measurable performance criteria is based on industry standards, such as the South Coast Air Quality Management District (SCAQMD) Rule No. 1168, which establishes VOC limits on adhesives and sealants. If no specific or measurable criteria for a green building product are included in the specification, then the contractor should either issue an RFI requesting clarification on the requirement or develop its own criteria for the RFQ.

Once the specific green building product performance requirements have been identified, the next step in the supplier RFQ process is to research commercially available products and identify one or more of those products that will meet the required performance requirements. This product research can be time consuming because of the specific performance requirements and certifications that green building products often have to meet. However, time spent now researching products and product performance data is a good investment, because this could avoid problems later during construction or when a third-party review is being conducted at the end of the project to certify the building's greenness. The outcome of this step in the supplier RFQ development process for performance specifications should be a set of acceptable green building products supported by manufacturer test data and certifications. If only one product is found that can meet the specification, the contractor should submit an RFI to ensure that this is what was intended when the specification was drafted, as well as offer voluntary alternates to try to get more than one supplier to quote for the required product.

The last step in the performance specification path is to prepare a product specification that will allow suppliers to provide a quote for the green building product. In the case of a custom building product that needs to be fabricated rather than mass produced, this would be a detailed performance specification. However, if standard off-the-shelf products can meet the performance specification, then the product specification may just be a list of acceptable products that suppliers can quote on.

Descriptive Specification Path. The descriptive specification path for a green building product is very much like a standard product that is specified using a descriptive specification, as illustrated in Figure 7-5. Typically, a descriptive specification describes the product wanted, including green requirements and a list of acceptable products. However, the contractor still needs to analyze the descriptive requirements to ensure that they are complete, up to date, and meet project-specific green building product requirements or any referenced third-party green building certification systems incorporated into the contract documents. If the descriptive requirements are not correct, then the contractor needs to issue an RFI to clarify the requirements.

If there are listed acceptable building products associated with the descriptive specification, the contractor needs to research these listed products to ensure that they meet the requirements of the narrative specification. It is possible that the narrative specification contains green building product characteristics such as those listed and discussed earlier in this chapter that cannot be met by the products listed as part of the specification because the manufacturer's product specifications changed or the designer failed to adequately investigate the products listed. In the event that any or all of the listed products do not meet the narrative portion of the descriptive specification, the contractor should issue an RFI to get clarification before submitting its proposal or bid.

Many descriptive specifications that include a list of acceptable products also include the phrase "or equal," allowing the contractor to propose alternate products that meet both the narrative portion of the descriptive specification and that are considered equal in quality to the listed products. Quality in this case typically refers to other product characteristics that are not easily measurable, such as manufacturer reputation, experience, reliability and maintainability, customer support, and other qualitative factors. Proposing an "or equal" product can sometimes help the contractor gain a competitive advantage in the bidding or proposal process as well as lower the cost of the project for the owner. As shown in Figure 7-5, if there is an "or equal" option, the contractor may want to research alternate green building products that can meet the specification.

Prescriptive Specification Path. A prescriptive specification differs from a descriptive specification in that the specification is written around only one acceptable product. A prescriptive specification can include a narrative describing the green building product requirements along with the one acceptable product, or it could just state the acceptable product. Like the descriptive specification, the contractor needs to determine if the building product specified will meet the project-specific green building product requirements or any referenced third-party green building certification systems incorporated into the contract documents. If the prescriptive requirements are not correct, then the contractor needs to issue an RFI to clarify the requirements.

Single-product specifications eliminate competition among suppliers and often result in higher prices for the owner. As a result, owners and designers will usually permit the contractor to submit a voluntary alternate with its bid that includes a building product that the contractor offers as an alternate for the named product. During the review of contractor bids, the owner and designer will consider the alternate product and its impact on project performance and cost. With the rapid growth in the green building market, products available to choose from are increasing. Just like suggesting an "or equal" product, the submission of a voluntary alternate to a single-product specification may provide the contractor with competitive advantage.

Identify Suppliers. All three specification paths lead to activity where building product suppliers are identified. In many cases, the suppliers may already be identified during the analysis and research activities in each of the three specification paths. However, most of the focus up to this point may have been on building product manufacturers, and now the contractor needs to identify potential suppliers, such as distributors and manufacturer representatives that the manufacturer sells to. The end result of this activity is a list of acceptable suppliers, which could include manufacturers, manufacturer representatives, distributors and supply houses, and others depending on the quantity and type of building product being purchased.

Prepare and Issue RFQ. The last step in the supplier RFQ process shown in Figure 7-5 is for the contractor to prepare and issue an RFQ for the green building product to the qualified suppliers. The end of this process links to the product procurement process shown in Figure 7-6 through the supplier quotation.

7.7.3 Product Procurement Process

The green product procurement process is illustrated in Figure 7-6. The procurement process for a green building product is very similar to that for a standard product. However, as noted throughout this chapter, it is

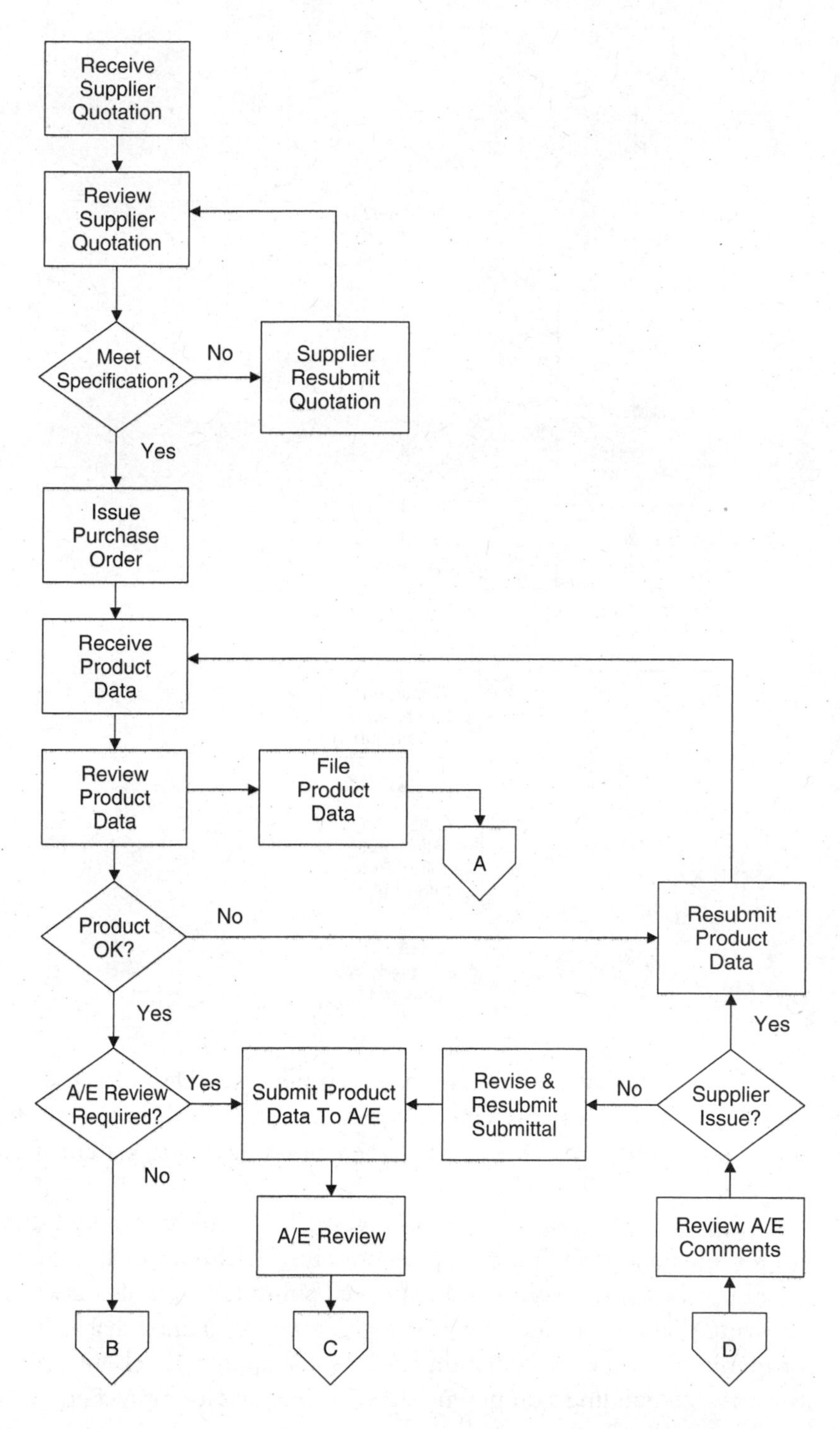

Figure 7-6 Product Procurement Process.

Figure 7-6 *(Continued)*

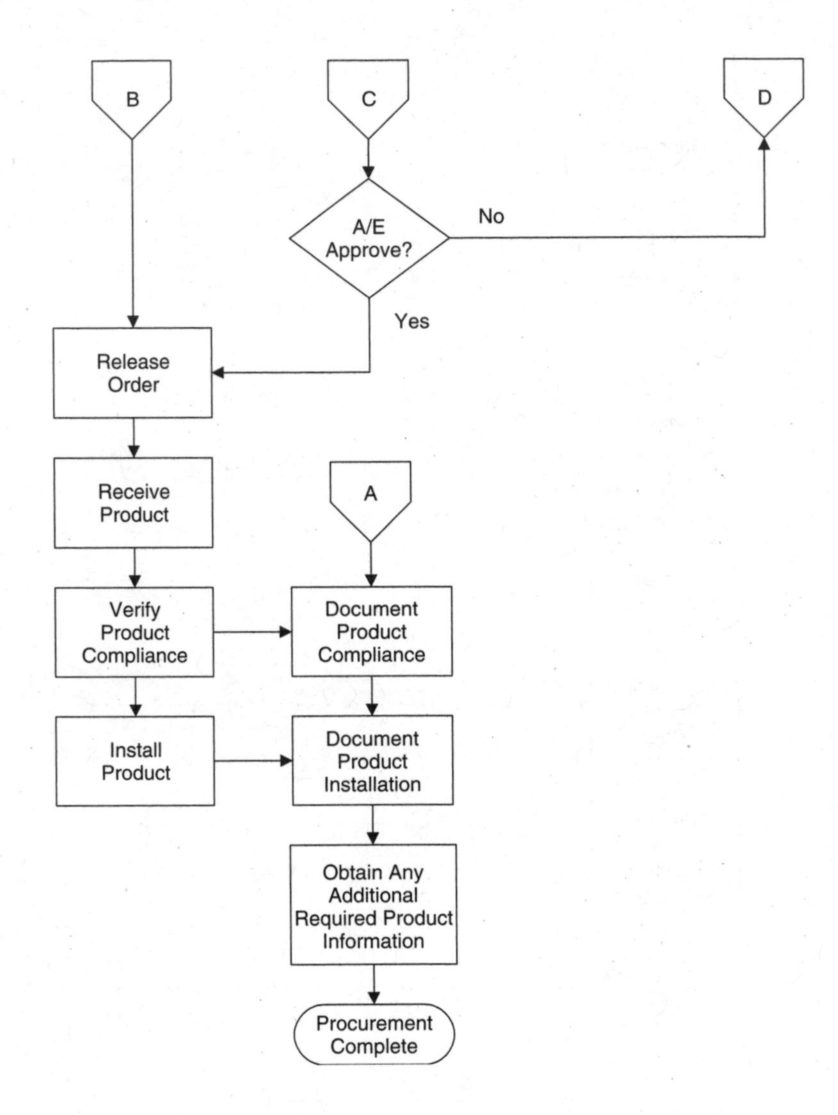

very important that the material or equipment purchased for use on a green building project meet the green requirements set forth in the contract documents as well as the third-party green building rating system used on the project.

As shown in Figure 7-6, maintaining a complete product file is very important on a green building project. This file should include not only the usual product information and approved submittals but also documentation showing that the product meets the green requirements, that it was installed properly, and other information such as startup and checkout procedures in the case of building equipment such as fans and pumps. This information may be required by the project contract documents or the green building rating system being used. Also, for building equipment, the information included in the product file will be needed for startup and commissioning as well as included in the operation and maintenance manuals (as discussed in Chapter 9).

7.7.4 Establish a Procurement Procedure

The contractor should establish a procurement procedure for green building projects to ensure that building products meet green requirements and to eliminate maverick purchases in the field. In addition, the contractor should make sure that subcontractors also have procurement procedures in place so that only acceptable building products are purchased and installed. These policies and procedures should be published and incorporated into subcontractor contracts where appropriate.

7.8 CASE STUDY

Creative Contractors, Inc.

City of Dunedin Community Center, Florida

In November 2006, Creative Contractors Inc., of Clearwater, Florida, completed the new, 45,00-square-foot LEED Silver Certified Community Center for the City of Dunedin, Florida. Designed by Collman & Karsky Architects of Tampa, it is the first environmentally friendly ("green") community center in the state of Florida.

Construction began in November 2006 with the demolition of the old, 20,000 square foot community center, more than 70 percent of which was recycled. The new Center is comprised of seventeen different rooms including a large multipurpose room with a stage and A/V equipment fit for full -cale stage productions. Other rooms include a state-of-the-art fitness center, dance and exercise rooms, a library and computer laboratory, an arts and crafts room, a recording studio, office space for Center employees, a kitchen, and classrooms. Additionally, the Center has a 1,500-square-foot outdoor performance stage and a state-of-the-art playground. The "Boundless Playground" is a handicapped-accessible playground that includes a soft floor made of 100 percent recycled rubber.

Figure 7-7 Photo Courtesy of George Cott.

Green Attributes

Green features include a single-ply, highly reflective, white, "cool roof" material for all roofs. The Energy Star-rated roof allows for less heat transfer into the building; the low-E windows provide similar protection for the outer walls. All of this provides for a very energy-efficient building envelope. Another "green" aspect of the building was the use of parking lot light fixtures that shielded light from spilling off the site into close residential neighborhoods and substantial grass parking area, decreasing the use of asphalt.

The building was designed to be 30 percent more efficient than a standard building; this equates to $35,000 per year in operational costs savings! The building design incorporated materials with recycled content manufactured regionally (within 500 miles of the jobsite.) Paints, adhesives, and sealants all had low-VOC (volatile organic compound) content to insure superior indoor air quality. Nearly 70 percent of all construction waste and demolition was recycled; that's more than 2 million pounds of waste diverted from the waste stream.

"Green" Facts and Figures

- 41.37% of materials were manufactured within 500 miles of jobsite
- 24.54% of materials were extracted within 500 miles of jobsite
- 11.27% of materials were made of recycled content
- 100% recycled rubber for playground floor

- Low-E windows (for greater insulation)
- "Green" housekeeping products will be used in the building
- Reduced use of pavement - more than 30% of the site consists of light-colored/high-albedo materials and grass reinforced with "Geo-Web" for traffic/parking
- 20% less water usage with low-flow plumbing fixtures

Lessons Learned:

Great diligence and follow-up is needed when collecting the backup data for the construction of a green building. Data for recycled and regional content as well as low-VOC documentation should be obtained as early in the process as possible, preferably during the submittals phase. Regardless of when you look for the data, it will often take multiple calls, emails, and faxes. For instance, we got to the end of construction of the Dunedin Community Center and needed the documentation for the low VOC carpet. The manufacturer's representative had told us all along that it met the LEED criteria. When went to get the product data sheets for the six walk-off carpets used at the building entries, the carpets appeared to *not* comply with the Carpet and Rug Institute Green Seal standard, which LEED requires.

The project team was ready to throw in the towel on the low-VOC carpet credit, because even though all other carpets in the building met the requirement, LEED does not leave room for even small walk-off carpets to be noncompliant. We called the rep, who assured us that the walk-off carpets must meet the LEED requirement, and he did not know why the product data sheets did not indicate them as such. He called his superior, who put us in touch with the company's environmental director, who found out that the product data sheets had not yet been updated but assured us the carpets did meet the LEED criteria. Problem solved! He got the data sheets updated, and the credit was salvaged. Had we not been diligent, the point would have been lost for no reason at all. It seems kind of trivial, but until the suppliers all get up to speed with green design and construction, it will be up to the contractors to be diligent and get the necessary data.

Similarly, the recycled content of our steel roof joists in our initial LEED submittal for the Dunedin Community Center was rejected. The reviewer said that the manufacturer's data showed that the 75 percent post-industrial content was created as part of their own manufacturing process and as such did not count toward recycled content for LEED. We called the manufacturer and explained the situation, and he felt bad but said that's just the way it is.

Upon further discussion internally, we knew that there's no way the steel being used is "virgin" or raw steel. We called the manufacturer back and explained further what we were looking for. A light bulb went off, and he said, "Oh! You need to speak with so-and-so." Well, we spoke with "so-and-so," and he informed us that the steel was 100 percent recycled content from various sources. Again, a credit saved because of contractor diligence.

The need for consistent communication of green goals to subcontractors and suppliers is essential for success. As such, we hold pre-bid meetings with subcontractors to discuss and educate them on LEED. Then, after subcontractor selection and before buy-out, we go over specific responsibilities with the subcontractors. Then, during construction, we bring up LEED at every subcontractor weekly meeting. This helps keep everyone focused. We also put signage on the job-site reminding everyone that this project is a bit different and requires special attention to recycling, indoor air quality, and so on.

7.9 REFERENCES

Bay Area Air Quality Management District, BAAQMD Rules & Regulations, www.baaqmd.gov/dst/regulations/index.htm, August 21, 2007.

The Carpet and Rug Institute, *Green Label/Green Label Plus*, www.carpet-rug.com/commercial-customers/green-building-and-the-environment/green-label-plus/index.cfm, August 21, 2007.

Forest Stewardship Council, *Principles & Criteria*, www.fscus.org/standards_criteria/, August 21, 2007.

Froeschle, Lynn M., Environmental Assessment and Specification of Green Materials, *The Construction Specifier*, October 1999, p. 53.

Green Seal, *Green Seal Standards & Certification*, www.greenseal.org/certification/standards.cfm, August 21, 2007

South Coast Air Quality Management District, *Rules & Regulations*, www.aqmd.gov/rules/, August 21, 2007.

The Sustainable Forestry Board (SFB) and American Forest & Paper Association (AF&PA), *Sustainable Forestry Initiative Standard (SFIS): 2005-2009 Standard*, 2004, www.sfiprogram.org/standard.cfm, October 25, 2007.

U.S. Environmental Protection Agency and the U.S. Department of Energy, Energy Star, www.energystar.gov/index.cfm?c=home.index, August 21, 2007.

chapter 8

Constructing a Green Project

8.1 INTRODUCTION

This chapter addresses constructing a green project and those aspects of sustainable construction that specifically impact the contractor's construction operations. Also covered in this chapter are the measurement and documentation requirements that may be imposed on the contractor during construction by the contract documents or third-party green building certification process. This chapter should also be helpful to the contracting firm that wants to be more proactive environmentally and incorporate green building methods into its day-to-day construction operations.

8.2 GREEN CONSTRUCTION PROCESS

Construction is a process, and green construction adds another dimension to that process. Figure 8-1 provides a diagram of the green construction process. Like any process, construction takes inputs, processes them, and then provides an output. In construction, the process inputs are labor, equipment, and material. The process is the construction means and methods employed by the contractor to transform labor, equipment, and materials into output. Output for the construction process is work in place at the construction site.

Green construction is also concerned about construction waste, as shown in Figure 8-1. Construction waste is defined as any portion of a construction input that is either not incorporated into the work or a byproduct of the construction process that is not incorporated into the work. The construction waste stream is more than just materials. Poor planning, poor work

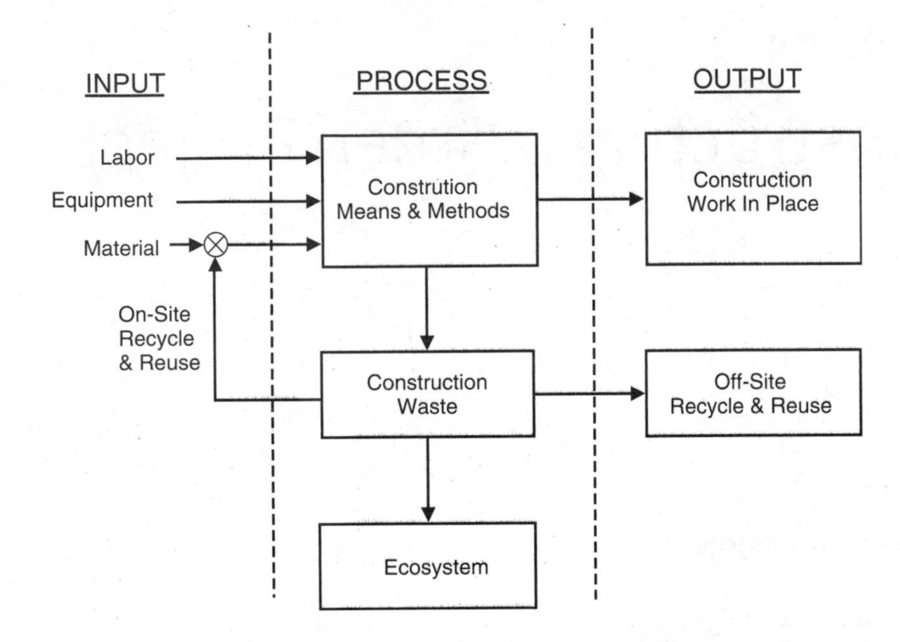

Figure 8-1 Green Construction Process.

environment, and accidents can also waste labor productivity and hours. Poor planning, untrained operators, unnecessary damage to the equipment, and pollution resulting from the equipment use can waste equipment productivity and hours. Green construction is about reducing this waste stream and improving the overall efficiency of the construction process.

The goal of green construction should be to eliminate as much waste as possible. Not producing waste is both the most efficient and cost-effective approach to sustainability. If construction waste has to be produced, it should be recycled or reused on-site, reducing the amount of new materials or off-site recycled and reused materials needed to be incorporated into the work. Where it is not possible to recycle or reuse construction waste on-site, the next best thing is to recycle or reuse it off-site in some manner. Wherever possible, green construction attempts to avoid returning the waste to the ecosystem where it is either landfilled, incinerated, or dumped into the atmosphere as pollution.

The efficiency of any process is determined by the amount of output obtained from a given set of inputs. The efficiency of the construction means and methods determines the amount of work in place for a given input of labor, equipment, and material. However, green construction is about sustainability through the conservation of resources, which represent the inputs to the process. Improved efficiency through reduced waste will improve both construction quality and contractor profits.

8.3 GREEN CONSTRUCTION PLANNING AND SCHEDULING

Construction planning and scheduling must account for the unique aspects of green construction. Sustainable project requirements need to be addressed in the construction plan and must be included in the schedule. In particular, green project requirements will impact the procurement, construction, and project closeout and commissioning.

Material procurement for sustainable construction projects, whether green materials or standard materials and equipment, can impact the sequencing of construction activities as well as activity timing. Project specifications and green building rating systems sometimes require that certain types or a certain percentage of materials be procured regionally or within a given radius around the project site. A large project may strain a regional supplier's ability to support the planned rate of construction. The contractor needs to make sure that local and regional material suppliers are capable of meeting the planned production schedule.

Another schedule consideration is the need to minimize site disturbance during construction or to protect materials and equipment from contamination during the construction process. From a delivery standpoint, both of these possible requirements affect the delivery schedule of materials and equipment and restrict the contractor's ability to stockpile materials to avoid shortages during construction. Laydown and storage areas may be nonexistent on a green project in order to minimize site disturbance. Restricted laydown and storage areas will require the contractor to approach the project in the same manner as a project in a congested downtown metropolitan area, where there is no on-site storage, even though the project may be located in a suburban or rural area. Similarly, porous materials such as drywall that can absorb moisture in storage and environmental system components such as ductwork may not be able to be stored on-site because of concerns about mold and other contaminants.

Restrictions on green project site disturbance and concern over contamination of materials and equipment stored on the construction site promote just-in-time (JIT) delivery of materials and equipment. As a result, the contractor's planning process must account for JIT material and equipment delivery, and closer coordination with material and equipment suppliers is necessary to ensure an adequate stream of materials and equipment to meet planned production rates. It is always good practice to include procurement activities in the construction schedule on traditional projects, but it is imperative to include procurement activities in the construction schedule on green projects. Also, the schedule detail associated with material and equipment procurement must also increase to provide an effective means of monitoring and controlling material and equipment deliveries.

During construction, green requirements and constraints can impact construction sequencing and timing. For instance, the installation of major HVAC duct runs is often begun in conventional buildings before the building is enclosed and weather tight. On a green building, starting installation of ductwork may not be allowed until the building is enclosed to avoid the possibility of contamination by dust and moisture. Changes in activity sequence and timing like this can impact the contractor's schedule and need to be planned for at the beginning of the project to avoid problems at the end.

Similarly, as will be discussed in Chapter 9, commissioning and closeout is much more involved and time consuming on a green building project than on a conventional building project. Commissioning and closeout activities need to be built into the project schedule, because they are critical to building completion and need to be monitored and controlled just like any procurement or construction activity.

8.4 WORKFORCE CONSERVATION

8.4.1 Need to Conserve the Workforce

Buildings are built by people. The shortage of skilled craftworkers is projected to worsen as many current workers retire and the construction industry finds itself competing with many other industries for the craftworkers of tomorrow. A skilled workforce in construction is a must, and this workforce must be cared for and preserved just like any other resource used in the construction industry. The contractor must recognize this situation and strive to conserve its skilled craftworkers as well as attract and train new workers to make up for retirements today and meet the needs of tomorrow's construction market.

8.4.2 Safety

First and foremost, the contractor must strive to provide a safe workplace for its workers and protect them from injury. A lost-time injury on the job site not only results in the temporary or permanent loss of a craftworker but also impacts the productivity of the crew with whom the injured worker was working, takes field management away from the task at hand and requires it to focus on the mishap and all of the required reporting that goes with it, as well as impacts the contractor's workers' compensation insurance premiums.

8.4.3 Ergonomics

The contractor's safety program typically focuses on acute injuries such as cuts and abrasions that manifest themselves when the accident happens and require immediate attention. Musculoskeletal disorders (MSDs) are also often treated as acute injuries and are focused on when they manifest themselves and the ability of a craftworker to work is affected. In fact, MSDs are not acute injuries and are usually the result of ongoing trauma over weeks, months, or years. MSDs are about injury to the soft body tissues of workers and include everything from carpal tunnel syndrome to lower back pain. To keep skilled and experienced craftworkers on the job, the contractor needs to address occupational ergonomics, which strives to match the workers' physical capabilities to their jobs through work design, personal protective measures, and other approaches.

8.4.4 Use of Local Workforce

Whenever possible, the contractor should use the local workforce, because this will reduce the commute to work, support the local community, and involve the local community in the project.

8.5 MATERIAL CONSERVATION

Green procurement is covered in Chapter 7, which focused on the various types and characteristics of green materials that may be incorporated into construction. The contractor needs to be aware of what green materials are, what makes them different from standard building materials, any industry standards that need to be considered when procuring green materials, and special considerations in procuring green materials. The procurement process is very important for the contractor, because it will determine whether the contractor gets materials with the specific green characteristics that it wants or is specified and, if applicable, meets the requirements of the green building rating system being used on the project.

This chapter focuses on the green construction process, and this section focuses specifically on waste reduction in the construction process. Material conservation applies not only to green materials but any materials that are brought on site. The material conservation methods discussed in this section will not only reduce the waste stream from the project but also potentially improve productivity, reduce direct construction costs, and improve the contractor's profit.

8.5.1 Material Conservation Planning

Material conservation planning should start with the bid preparation process and continue through the preconstruction process. Material conservation planning should be a team effort, including the designer, if the contractor is involved in the planning and design stages of the project, as well as material manufacturers and suppliers. Where materials will be purchased and installed by subcontractors, these subcontractors should be encouraged to procure their materials and plan their work to minimize material waste, which may require education and training by the contractor (as discussed in Chapter 6). Material conservation planning should be an important part of any constructability reviews or value analyses performed by the construction team during the planning and design process if involved and during bid preparation and preconstruction planning.

8.5.2 Material Conservation Strategies

Many material conservation techniques can be employed by the contractor to reduce material waste. Some common material conservation strategies are as follows:

- Design to standard material dimensions.
- Fabricate nonstandard material dimensions.
- Prefabricate material assemblies.
- Evaluate material shipping.

The successful application of these and other material strategies will depend on the project characteristics that include the size of the project and the volume of material that is involved. For a small project or a small part of a larger project, it might not be economical to employ these strategies. However, when considering employing any material conservation strategy, the contractor should consider the overall impact on the project and the project goals regarding waste management and minimization. Reducing the project waste stream to meet specification requirements or fulfill a green building rating system requirement may make the use of a project waste management strategy necessary, even though taken by itself it would not be economical.

Build with Standard Material Dimensions. Building with standard material dimensions reduces not only waste but should also reduce material costs through reduced waste and improved field productivity, because field crews do not have to fabricate materials. To a large extent, the ability to build with standard material dimensions has its roots in the building planning and design. Building dimensions are set during the early stages of design, and it is easiest to consider standard material dimensions at this time. If the contractor is involved with the project during the planning and design, design reviews are an excellent time to identify places where changes in building interior or exterior dimensions or shape can facilitate the material installation process. As always, the cost of making a change as well as the impact on other building systems needs to be considered. For example, changing the building floor-to-ceiling height to standard material dimensions may not be economical. However, making a slight change in layout, such as repositioning a doorway that will be replicated many times over as in the case of a hotel, can result in a large savings of both materials and labor.

If the contractor is not involved in the planning and design of the building, a voluntary alternate included in its bid or a value analysis proposal after contract award may be a way to address the use of standard material dimensions. If the cost savings are large enough or if the proposed change will reduce waste and improve productivity significantly, it may be worth the cost of redesign to the project. For example, repositioning roof-mounted equipment and other roof penetrations may not only reduce roofing material waste and improve roofing crew productivity but may also reduce the number of membrane seams and penetrations, which will reduce the probability of roof leaks and increase the life of the roof.

Building to standard material dimensions can be extended to project specialty contractors and suppliers that fabricate materials for incorporation into the work. For example, most sheet metal fabrication firms that build HVAC ductwork have their own shop standards that are best for their fabrication facility and equipment. Others have standardized on industry standards, such as the Sheet Metal and Air Conditioning Contractors' National Association, Inc. (SMACNA) *HVAC Duct Construction Standards: Metal and*

Flexible, as a basis for shop fabrication in the absence of specific construction contract requirements.

Similarly, sprinkler contractors often fabricate sprinkler piping off-site at their fabrication shop based on the project specification requirements. By allowing specialty contractors to select materials, shapes, dimensions, and other characteristics of fabricated materials that best match their fabrication process and do not compromise the performance of the system, waste in the fabrication and installation process can often be reduced and productivity increased.

The fabrication of temporary construction systems and structures can also reduce material use and improve productivity. An example of this would be formwork. Designing the building to use standard formwork for concrete placement will reduce the number of forms required, whether they are metal, wood, or other material. This will reduce materials, which may be especially important where there is a limited amount of concrete work or concrete details that require custom on-site wood form construction. Using standard formwork sections will also reduce the forming crew's cycle time, shorten the contractor's pour schedule, and possibly shorten the overall project schedule, because concrete placement is usually on the critical path, especially for a concrete frame building.

Fabricate Nonstandard Material Dimensions. If it is not possible to build with standard material dimensions, the next best thing is to fabricate materials to fit the project where the quantity of materials is sufficient to warrant any additional costs that may result from fabricating materials to nonstandard material dimensions. Fabricating materials with nonstandard dimensions requires a careful analysis of the project to determine not only what those dimensions are but also the impact on other related building systems, material shipping to and handling at the project site, and the impact on installation crew productivity, among other considerations.

An example of this occurred on a large multistory building expansion that included a lot of floor duct for power and communications in open office areas. The building was laid out with 24-foot runs of floor duct as a result of column spacing. The standard manufactured length of floor duct is 10 feet, and the use of standard material would have required hundreds of field cuts, which would have resulted in reduced productivity and waste. The electrical contractor worked with its supplier and was able to have the floor duct manufactured in 12-foot lengths as well as any custom lengths that would be needed. This eliminated all field cuts and not only improved the electrical contracting firm's productivity and eliminated waste but also expedited the contractor's pour schedule, which improved the efficiency of the concrete placement on each floor and reduced the overall project schedule.

Prefabricate Material Assemblies. Material assemblies can be prefabricated off-site under controlled conditions to reduce waste at the jobsite as well as avoid problems that could occur when certain materials are required to be used. For example, prefabricating audio cable assemblies for a theatre will reduce the need to pull and terminate individual cables, which in turn will reduce waste and improve the sound contractor's installation productivity. Similarly, prefabrication off-site can avoid conflicts between what is needed to meet the contract requirements and provide a quality installation and the requirements of the green building rating systems being used on a project. For example, a solvent-based adhesive or paint may need to be used for durability or other reasons, but this solvent-based material may emit volatile organic compounds (VOCs). If applied on-site, these VOCs would need to be addressed in the construction IAQ program, which may increase the cost of the operation and lower productivity. Instead, the solvent-based material or paint could be applied in the manufacturer's facility or specialty contractor's shop under safe and controlled conditions and avoid the need to address the VOCs on site.

Evaluate Material Shipping. Waste can occur in shipping as a result of material breakage and spoilage. When planning material deliveries, the contractor should consider the mode of shipping and its impact on material breakage and spoilage as well as the cost of shipping and installation. For example, the contractor constructing a building with a granite curtain wall noted that the quality of granite panels and breakage during shipping could be greatly decreased if the architect would allow the use of panels one-fourth the size specified. The change in panel size did not noticeably impact the building aesthetics, but material waste caused by breakage both during shipping and construction was significantly reduced. In addition, the use of smaller pieces of granite allowed the use of what otherwise would have been scraps with the original panel size.

8.6 SITE LAYOUT AND USE

Construction site layout and use is very important on any construction project but it is especially important on green building projects. One of the goals of green construction should be to reduce the disturbance to the site during construction in order to preserve the natural setting and habitat around the building. If the site has been previously disturbed, the goal should be to minimize further disturbance and damage to the site and, to the extent possible, return the site to its natural state following construction. Reducing site disturbance not only addresses the building site but also protects surrounding land, both developed and undeveloped, from disturbance or damage during construction. Third-party green building rating systems typically have specific requirements for minimizing site disturbance during construction, and the contractor should be familiar with those specific requirements if credit is sought for minimizing site disturbance. Strategies that the contractor can use to ensure reduced site disturbance include the following:

- Set construction boundaries.
- Restrict vehicle and equipment movement.
- Establish trailer, storage, and laydown areas.
- Prevent site erosion and sediment runoff.
- Manage stormwater and wastewater.

8.6.1 Set Construction Boundaries

In order to prevent site disturbance during construction, the contractor needs to set and mark boundaries around the perimeter of the building, as well as around hard surfaces such as surface parking areas, sidewalks, driveways, and access roads that are part of the construction project. In addition, corridors should be established for incoming utility lines to prevent damage. The boundaries should be clearly marked and all subcontractors informed of the boundaries within which they will need to work. Subcontractors should be notified of these restrictions before bidding, because these boundaries may restrict the equipment they could use and impact their productivity, which will need to be addressed in their bid.

8.6.2 Restrict Vehicle and Equipment Movement

Along with setting construction boundaries, vehicle and equipment movement on the site should be restricted to prevent site disturbance. These boundaries may be the same as those around the building perimeter and hard surfaces, or they may be different for specific equipment to allow its use. If boundaries are extended for specific equipment or construction operations, the contractor should have a plan for restoring the affected area to its natural state. Also, if the contractor is going to make a subcontractor responsible for restoring the site after performing a construction operation outside of the established boundaries, this requirement should be communicated to the subcontractor before bidding and included in the subcontract requirements.

8.6.3 Establish Trailer, Storage, and Laydown Areas

Trailer, storage, and laydown areas should be established at the site before mobilization. The contractor should work with subcontractors before mobilization to understand their trailer, storage, and laydown area needs throughout the construction process. As a result of the need to minimize site disturbance and limited area with which to work, the contractor may have to restrict the number of trailers that subcontractors are allowed to move on-site as well as restrict storage and laydown areas. Subcontractors may need to move equipment off-site when it is not in use when they would have normally just left it parked on-site until it was needed again. Similarly, storage and laydown areas may be limited and need to be shared by subcontractors as needed during the project.

Once work is complete on a particular phase of a project, excess materials may need to be moved off-site immediately to make room for the next subcontractor's materials. All of this will require the contractor to work closely with subcontractors throughout the construction process to maintain the site and accommodate their subcontractor storage needs for current construction activities.

8.6.4 Prevent Site Erosion and Sediment Runoff

The contractor should implement measures to prevent site erosion and sediment runoff during construction. The exact measures taken to prevent site erosion and sediment runoff on a particular green building project will depend on soil conditions, site topography, vegetation, and other physical site characteristics as well as contractual, third-party green building rating system, and applicable federal, state, or local government requirements.

8.6.5 Manage Stormwater and Wastewater

Similarly, the contractor should take steps to manage stormwater and wastewater during construction so that runoff does not pollute or damage the building site or surrounding areas.

8.7 CONSTRUCTION WASTE MANAGEMENT

As discussed earlier in this chapter, one of the goals of green building construction is to minimize construction waste. The best way to accomplish this goal is to first eliminate waste wherever possible by ordering only materials that are necessary, minimizing packing and shipping materials, using standard-size building products wherever possible, and custom-fabricating products where standard-size building products cannot be used. However, it is not possible to eliminate all construction waste, and the contractor needs to develop and implement a waste management plan at the construction site aimed at reducing the amount of waste that is returned to the ecosystem via landfills and incineration. As shown in Figure 8-1, the goal is to recycle or reuse as much construction waste as possible either on-site or off-site.

The contractor needs to first establish a measurable goal for the reduction of construction waste to be disposed. Third-party green building rating systems typically award credit toward building certification or verification as a green building based on the amount of waste that is diverted from landfills and incineration using either percent volume or weight as the measurement. If a third-party green building rating system is being used on a building project, the contractor should understand how the rating system accounts for waste diversion and what the percent thresholds are for getting credit.

Working with subcontractors, the contractor can then determine what types of waste are expected on the project and a realistic percentage of waste that can be recycled or reused. The percent of project waste that can be recycled on the project depends not only on the amount of waste that the contractor believes will be generated and can be captured at the site, but also what recycling capabilities in terms of both materials and volumes that local waste management companies can handle. Therefore, in addition to working with subcontractors to set a realistic goal for construction waste reduction, the contractor also needs to work with local waste management firms to determine their capabilities to accept the waste.

Once a realistic goal has been set, an on-site waste management plan needs to be developed. This waste management plan should again be developed with input from both subcontractors and the local waste management company or companies that the contractor decides to contract with. For a large project with significant volumes of waste, the contractor may want to work with several waste management firms, particularly if they specialize or have capabilities for handling specific types of construction waste. For a small project, there will probably be only one waste management company.

The on-site waste management plan will largely depend on the capabilities and services that will be provided by the waste management firm. In most cases, waste will need to be segregated at the project site before pickup, which means multiple bins or Dumpsters to accommodate different materials. If it is not possible to have multiple bins or Dumpsters in a particular area or on-site, then consideration should be given to phased waste management, where bins or Dumpsters are used to collect recyclable waste from construction operations going on at the time and, when complete, the bins are reassigned for waste from the next activity. Collection is not as efficient with phased waste management, because other recyclable waste may not be able to be captured. With proper planning, most of the recyclable waste being generated at any one time should be able to be captured.

In order to ensure success, subcontractors and their field personnel need to understand the importance of waste management to the success of the green building project and what is expected of them. The contractor needs to communicate the project waste management plan to subcontractors, along with the procedures for recycling and reusing construction waste. Wherever possible, the contractor should make it as easy as possible for subcontractors and their field personnel to comply with the waste management plan. This includes placing signs and posters around the site to remind workers to use the recycling bins and Dumpsters instead of just throwing construction waste into general trash Dumpsters. If there are separate bins or Dumpsters on-site for different types of recyclable materials, make it easier for workers to know which is which by using easy-to-read bilingual signs where appropriate, color-coding Dumpsters, or providing another means of identification. Also locate bins and Dumpsters as close to the work area as possible to reduce the distance and time it takes for workers to dispose of recyclable waste properly.

8.8 MATERIAL STORAGE AND PROTECTION

Material storage and protection is very important on green building projects. Materials and equipment, particularly interior finish materials and HVAC and plumbing materials, must be protected to prevent contamination by dust, moisture, dirt, and mold before installation. If possible, materials should be stored off-site in a controlled environment until they are needed for installation. When brought on-site and stored while waiting to be installed, materials should be covered, sealed, and protected from damage and the elements. For instance, fabricated ductwork and piping should have their ends sealed when brought on-site, and the seals should only be removed when the materials are being installed. At the end of the day or if there is a break in an activity, any openings in piping or ductwork should be sealed until work resumes. Porous materials and fabric-based materials should also be kept dry and in a controlled environment until they are ready for installation.

8.9 PROVIDING A HEALTHY WORK ENVIRONMENT

Green construction is not just about providing a healthy work environment for building occupants after construction. Green construction is also concerned about providing a healthy environment for construction workers and the public during construction. During the demolition and deconstruction of existing structures, earthwork operations, and other outdoor site work, the contractor should control dust and other airborne pollutants as much as possible. For example, dust can be minimized during earthwork operations by keeping the ground damp using water trucks or sprinklers. Similarly, replacing older, heavy equipment with newer, more efficient and cleaner equipment when it is economically justifiable can reduce fuel use and minimize exhaust emissions. This will reduce pollution and improve the environment both on-site for workers and in the surrounding areas for the public.

Once the building is enclosed or nearly enclosed to the point that work can begin inside, the contractor needs to consider the interior environment in which the workers will be working. Of primary concern is the indoor air quality (IAQ), because of the dust that results from many building finish activities, such as drywall and finish carpentry, and the myriad of different chemicals in the form of adhesives, sealants, paints, and coatings that are used in the construction process. The contractor needs to develop and implement an IAQ plan during construction to ensure the health and well-being of workers. In addition, having an IAQ plan that is implemented during construction may be a requirement of the third-party green building rating system.

The guidelines that are typically cited in specifications and third-party green building rating systems for maintaining IAQ during construction are the *IAQ Guidelines for Occupied Buildings under Construction* (*IAQ Guidelines*). The *IAQ Guidelines* are published by the Sheet Metal and Air Conditioning Contractors' National Association, Inc. (SMACNA 1995). The objective of the SMACNA guidelines is to protect construction workers and building occupants from airborne dust, odors, and other harmful contaminants during demolition, construction, and punchlist activities. This publication is divided into seven chapters that address air pollutants associated with construction, control measures, managing the construction process, quality control, and other topics.

The key chapter in the *IAQ Guidelines* is Chapter 3, which covers control measures. These control measures are what is most often cited in specifications and third-party green building rating systems and what the contractor is required to meet or exceed. The five control measures addressed in Chapter 3 of SMACNA's *IAQ Guidelines* are as follows:

- HVAC protection
- Source control
- Pathway interruption
- Housekeeping
- Scheduling

HVAC protection addresses methods to prevent dust from getting into the air-distribution system as well as odors that can be absorbed by porous parts of the HVAC systems and released into the air stream at a later time. The most effective method of controlling pollution is at the source, and the guidelines for source control address methods for containing or eliminating air pollution at the source. If the pollutants cannot be controlled at the source, the *IAQ Guidelines* provide recommendations for pathway interruption, which is simply the prevention of air movement along with the migration of dust, contaminants, and odors through the use of barriers and pressurizing spaces. Housekeeping simply refers to keeping the indoor construction site as clean as possible through dust collection and cleaning. Scheduling involves carrying out construction operations that result in a lot of dust and contaminants at times when the building is not occupied and there is sufficient time for the air to clear before other workers and occupants return.

In order to condition the workspace for workers and materials during construction, the early startup and use of the building HVAC system is sometimes considered. This should be avoided if at all possible. First, the warranty period on HVAC equipment typically commences when the equipment is started up and will probably require that the HVAC contractor purchase an extended warranty from equipment manufacturers to provide the contractually required one-year warranty on the building and systems after substantial completion. Also, to prevent the buildup of dust and contaminants in the air-distribution system, filters with a Minimum Efficiency Reporting Value (MERV) of 8 are required to be installed at each return grille.

Furthermore, before occupancy, all HVAC system filters will need to be replaced and ductwork may need to be cleaned, which may not have been included in the HVAC subcontractor's bid price. For additional information on the early startup of the building HVAC system during construction, the reader is directed to SMACNA's position paper entitled *Early Start-Up of Permanently Installed HVAC Systems*.

8.10 CREATING AN ENVIRONMENTALLY FRIENDLY JOBSITE

The contractor can do several things to create an environmentally friendly jobsite beyond those requirements mandated by the contract documents and third-party green building rating systems. As discussed in Chapter 1, being green is becoming a way of doing business for many contractors, and they are benefiting from more efficient operations, reduced energy costs, and high employee morale. Some of the things the contractor can do at the jobsite include the following:

- *Establish guidelines for light levels and lighting quality for construction activities.* Construction activities require lighting outdoors both in the evening and the early morning. Similarly, indoor work away from windows and other sources of natural light requires temporary artificial lighting before the permanent building lighting is installed and operable. The contractor should consider establishing guidelines for lighting levels and quality during construction both outdoors and indoors. Proper lighting will improve the quality, productivity, and safety of the project.
- *Use renewable energy technologies for construction power.* This is already being done in the highway construction industry, where electric signs are being powered by photovoltaic (PV) panels. On a building site, PV panels could be used to charge batteries, which in turn could be used for security lighting at night.
- *Select and use energy-efficient light sources.* Most construction sites still use incandescent lighting for temporary lighting. Incandescent lighting equipment is inexpensive and easy to install, but it is inefficient and does not always provide the best light for the task. The contractor should investigate the use of other more efficient light sources at the project site, such as compact fluorescent lamps, especially if the lighting will also be used as security lighting and left on all night.

- *Use motion sensors for security lighting.* Security lighting on construction sites is normally left on all night. Some areas need to be lit all night for security and safety reasons, but many areas do not necessarily have to be continuously lit all night. The contractor should consider the use of motion sensors on security lighting to reduce energy use and expenses.
- *Avoid light pollution and protect the night sky.* Whether night lighting at the construction site is being used for security or construction operations, the contractor should consider the impact of construction site lighting on neighbors and the night sky. The contractor should consider setting and aiming light fixtures to avoid light pollution and protect the night sky. In addition, the contractor should consider using fixtures with a high cutoff where appropriate for the task at hand to further reduce light pollution.
- *Establish lighting control zones based on construction sequencing.* All of the lighting at a building construction site is often turned on whether there is activity in an area or not. The contractor should consider installing light switches for temporary lighting so that the lights can be turned off when no work is being performed in the area or where the daylight entering the work area is sufficient and the temporary lighting is not needed.
- *Collect and use rainwater and greywater.* The contractor should consider collecting and using rainwater and greywater in construction and cleanup activities as well as for watering site vegetations and dust control when potable water is not required.

8.11 CONSTRUCTION EQUIPMENT SELECTION AND OPERATION

The contractor can implement several strategies to reduce fuel consumption and pollution resulting from the use of construction equipment and vehicles at the site. These strategies are not only environmentally friendly but will also save the contractor money and increase productivity. Strategies that the contractor might consider to improve the environment, increase productivity, and reduce costs are as follows:

- *Plan equipment spreads to minimize cycle time.* Select types and quantities of equipment to minimize cycle time, which in turn will increase productivity and reduce fuel costs. It may be more cost effective on a project to rent a piece of equipment that is better suited to the work requirements than to use equipment that the contractor already owns or leases. If there is significant earthwork on a green building project, it may be worthwhile for the contractor to analyze the operation in greater detail than usual and select the type and quantity of equipment that would be best for the job. Equipment dealers often have or have access to simulation software that can help the contractor select the best equipment spread.

- *Train operators to use equipment efficiently.* Knowing how to operate a piece of equipment does not mean that an operator knows how to operate a piece of equipment efficiently. The contractor should consider training equipment operators on how to operate equipment efficiently.
- *Avoid unnecessary equipment idling.* The contractor should develop guidelines for equipment operators that include a maximum idling time when the equipment will be shut down if the operator thinks that he or she will exceed it.
- *Use battery-powered vehicles and equipment.* The contractor should consider substituting battery-powered vehicles like golf carts that can be recharged overnight for pickup trucks and other vehicles on-site where possible.
- *Replace gasoline and diesel with alternative fuels.* If available and compatible with engine requirements, the contractor should consider using alternative fuels in its vehicle and truck fleets.
- *Encourage worker use of public transportation and car pool.* Workers should be encouraged to use public transportation or car pool to conserve fossil fuels and reduce emissions. The contractor can do this by setting up a bulletin board on-site so that workers who live in the same area and are interested in carpooling can contact one another. In addition, the contractor can post routes and schedules for public transportation around the jobsite to generate interest and encourage the use of public transit.

8.12 DOCUMENTING GREEN CONSTRUCTION

A green construction project usually requires that the contractor prepare and provide additional submittals before the start of construction, during construction, and at project closeout. These submittals vary from project to project. It is very important that the contractor know what submittals are required, when they are due, what their format should be, and their review and approval process.

Green project submittals can include plans such as the following that need to be submitted, approved, and documented for compliance:

- Site Preservation and Use Plan
- Waste Management and Recycling Plan
- Indoor Air Quality Plan
- Material Delivery and Protection Plan

Similarly, submittals such as the following that detail the type and quantity of the following material categories may also be required:

- Salvaged and Refurbished Materials
- Recycled Content Materials
- Regional Materials
- Certified Wood Products

On green construction projects, product information and certifications in addition to the requirements in the technical specifications may also be required. This product information and certification submittal requirement is often in addition to the usual shop drawing process and may even be redundant. The purpose of these submittals is to demonstrate that green project requirements have been met and the extent to which they have been met. These submittals are often included as part of an application for certification as a green building by an outside third party. Some typical product information and certifications that may be required on a green building are as follows:

- Roofing
- Lighting Cutoff
- Plumbing Fixture Water Use
- HVAC Equipment without CFC Refrigerant
- HCFC Refrigerant

The contractor needs to know what green construction documentation it needs to supply. Requirements for green construction documentation will normally be found in the project contract documents and in any third-party green building certification system used on the project.

8.13 CASE STUDY

Oscar J. Boldt Construction

Wisconsin River Valley Office and Warehouse, Wisconsin

Overview

- Location: Stevens Point, Wisconsin
- Building Type(s): Commercial Office Building
- New Construction
- 23,000 sq. feet
- Project Scope: Single Story
- Urban Setting
- Completed September 2002
- Rating System: LEED®-NC v2.0
- Rating Level: Silver
- Credits Achieved: 33 credits

The Oscar J. Boldt Construction Wisconsin River Valley Regional Office is a 12,000-square-foot commercial office and an 11,000-square-foot warehouse and shop facility and has a 31,000-square-foot outdoor storage yard. The building is located in the high-profile Portage County Business Park along one of the busiest thoroughfares (Highway 51) in Wisconsin. It is Boldt's first experience of designing, building, documenting, and certifying a LEED®-NC project.

One of the interesting features of this facility is the concept of "Boldt gives back to the community" by allowing the community to utilize the state-of-the-art conference room for meeting space. Simultaneously, the facility offers the opportunity to allow the community to learn about sustainable principles and methods that reduce negative impacts on the environment and enhance occupant health and well-being. By allowing people to "kick the tires," it is easy to demonstrate what a "green" building is, what recycled materials look like, what it feels like to be in a naturally daylit space, how it feels to be in a space with high indoor air quality, and how to minimize energy consumption.

Figure 8-2 Photo courtesy of The Boldt Company.

Environmental Aspects

The project encourages alternative transportation through preferred parking for carpooling and hybrid vehicles and provides a bicycle rack and shower facilities for cycle enthusiasts. The site also focuses on stormwater management, light pollution reduction and utilized nonpotable water from the retention pond for landscape irrigation.

During construction, the project team implemented a construction waste management plan, which enabled the on-site personnel to recycle 79 percent of construction waste. The Boldt team continues to recycle now that they have occupied the building. The project team also specified materials with a high recycle content, materials that were assembled, manufactured, and harvested locally and purchased and installed FSC millwork and interior wood doors.

Figure 8-3 Photo courtesy of The Boldt Company.

A construction IAQ management plan was created and implemented during construction to ensure the health and well-being of on-site construction field personnel and the building was flushed out prior to occupancy. The project also features low-emitting materials such as adhesives, sealants, paints, carpet systems, and urea-free formaldehyde composite wood. The CO_2 level is monitored and automatically controlled in the conference spaces to ensure that the indoor air quality levels are maintained at a high level.

The facility focuses on water efficiency through stormwater management, high-efficiency irrigation from non-potable water drawn from the on-site retention pond, and the installation of low-flow plumbing fixtures.

Figure 8-4 Photo courtesy of The Boldt Company.

To meet the energy-efficiency goal of 40 percent better than the ASHRAE 90.1-1999 standard, the facility integrated natural daylight strategies, including the installation of specially tuned glazing that is shaded by large overhangs to reduce the solar load. More than 85 percent of the building is naturally daylit and provides direct views to the exterior for 100 percent of the employees and visitors. Other strategies include daylight and occupancy sensors, increased R-value of the building envelope, and high-efficiency lighting and mechanical systems that are monitored by a building energy management system. To ensure the energy efficiency, the project included fundamental and enhanced building energy-systems commissioning.

Environmental Benefits

There are many reasons why building owners want to incorporate green into their facilities. First, the green benefits that this project realized was energy savings. The actual energy savings measured after one year of building operations yielded a savings of 58 percent. The premium of the estimated 40 percent more energy efficient HVAC&R system was $35,000. If an HVAC system designed to the ASHRAE 90.1-1999 standards was installed, the DOE-2 model estimated that $54,268 would be spent annually basis for energy. The first-year annual energy costs were only 4 percent of the building's total operating costs, which resulted in an annual savings of $31,881, or an avoidance of $956,430 over the estimated 30-year life of the building; hence had a payback period of 1.1 years. In addition to the benefit of direct cost savings by reducing energy consumption, the benefit of lessening the negative impact on the environment by reducing carbon emissions is realized.

Lessons Learned

Throughout the construction of a project, the project management team and the on-site field personnel have three main responsibilities: quality, schedule, and budget. On a green project, these three responsibilities do not change; they integrate additional "green" tasks.

It is important that the construction team understands that they play a very important role in greening a building project. In fact, the construction team is responsible for implementing and documenting approximately one-third of the LEED®-NC credits.

By making the construction team aware of the green project goals and the LEED®-NC credits being sought, which includes clearly communicating the credit intents, requirements, and documentation and submittal requirements, the project will be successful. Communication and education are essential to ensuring green project success.

The additional tasks that must be incorporated into the traditional construction responsibilities of a project include:

- *Project kick-off meeting.* During the traditional project kick-off meeting, the construction manager and on-site superintendent must clearly convey the green goals of the projects, the "plans" that will be implemented, how to implement the plan, and the roles and responsibilities of all on-site field personnel.
- *Project schedule.* The project schedule must include additional "green" tasks that are not traditional, such as building flush-out. Any long-lead "green" materials must also be identified and incorporated to ensure on-time delivery. Lean construction is an effective tool that has been proven to eliminate waste through team collaboration, enabling the team to communicate, which aids avoiding material deliveries and workforce management issues.

- *Documenting the "Construction" LEED®-NC credits.* It is highly encouraged that all subcontractors provide an updated breakdown of their material costs and related information with their monthly pay applications. It is strongly advised that the General Contractor/Construction Manager update all of the LEED® credit information on a monthly basis to ensure adequate documentation for the LEED® submittal process.
- *Construction activity pollution prevention.* The project team must ensure the traditional erosion and sediment control plan incorporates the Construction Activity Pollution Prevention Plan requirements and the plan is provided to all on-site field personnel. The plan must not only be effectively implemented, but also documented via photos and weekly reports. If there are site disturbance limits, these also need to be identified and explained to the on-site field personnel. It is encouraged to review the CAPP Plan with the on-site field personnel during traditional "Tool Box Talk" meetings.
- *Construction waste management.* The project team must effectively implement and track the recycling of construction waste. The team must ensure all on-site field personnel are aware that there are recycling goals, and a CWM Plan must be provided to all on-site field personnel, who must have a clear understanding of how the plan will be implemented. It is encouraged to review the CWM Plan with on-site field personnel during traditional "Tool Box Talk" meetings.
- *Materials.* The project team must ensure the green materials that are specified and approved in the submittal process are purchased and installed. The most difficult aspect of this is making sure that all on-site field personnel are aware of "green" materials such as low-emitting adhesives and sealants, urea-free formaldehyde composite wood, and FSC wood are purchased and installed. These materials should be clearly identified or displayed on posters in the job trailer or at the on-site field personnel break areas. It is also a good idea to make "green talk" a part of the traditional Tool Box Talks.
- *Indoor air quality.* Making sure that all on-site field personnel have a copy of the IAQ Management Plan During Construction and that the specific tasks required in the plan are implemented by all on-site field personnel. This includes the protection of all ductwork and the storage and protection of all absorptive materials. Also making sure that the

on-site field personnel are aware of low-emitting material VOC limits and that *all* of the adhesives and sealants being used on the inside of the vapor barrier meet these requirements. It is advised to post the VOC limits in the job trailer or at the on-site field personnel break areas, as well as review the VOC limits in the traditional "Tool Box Talks."

The Boldt Company

The facility is owned by BBO LLC and operated by Oscar J. Boldt Construction, Wisconsin's largest general contractor, a division of The Boldt Company, a 117-year-old construction company, headquartered in Appleton, Wisconsin.

8.14 REFERENCES

Sheet Metal and Air Conditioning Contractors' National Association, Inc. (SMACNA), *IAQ Guidelines for Occupied Buildings under Construction*, First Edition, 1995.

Sheet Metal and Air Conditioning Contractors' National Association, Inc. (SMACNA), *Early Start-Up of Permanently Installed HVAC Systems*, SMACNA Position Paper, Undated.

chapter 9

Green Project Commissioning and Closeout

9.1 INTRODUCTION

Commissioning and closeout of a green construction project is more complex than a traditional building project, particularly if the owner is seeking third-party certification. The purpose of this chapter is to help the contractor understand commissioning and closeout requirements and process, so that this work can be properly incorporated into the contractor's bid or proposal and efficiently and effectively performed. This chapter addresses green building commissioning, including understanding the contractual requirements for commissioning, the need for a comprehensive mutually agreed-upon commissioning plan early in the project, working with an outside owner-appointed commissioning authority, typical requirements for system startup and testing, and typical documentation that needs to be submitted. In addition, typical contract closeout requirements for green buildings are also addressed in this chapter, including submission of project documentation such as record drawings, addressing warranties and guarantees required by the contract documents, and training the owner's personnel.

9.2 BUILDING COMMISSIONING PURPOSE AND OBJECTIVES

The purpose of building commissioning is to ensure that building systems operate and can be maintained in accordance with the owner's project requirements as expressed in the facility program that defines the owner's operational requirements. The purpose of commissioning is accomplished by achieving the following four commissioning objectives:

- Verify and document that the equipment constituting building systems to be commissioned has been installed properly and operates correctly based on predefined procedures that include inspection, testing, and startup.
- Verify and document that each building system being commissioned operates correctly and interacts as required with other building systems.
- Ensure that complete equipment and system documentation, including operation and maintenance (O&M) information, is provided to the owner at the end of the project in an organized manner that can be easily accessed and used by the owner's operating personnel.
- Ensure that the owner's operating personnel have been properly trained in the operation and maintenance of commissioned building systems in order to ensure reliable, efficient, and sustainable building operation.

9.3 BUILDING COMMISSIONING DEFINED

Building commissioning can be defined as follows:

> Building commissioning is the systematic process used to verify that the completed building and the systems that comprise it operate in accordance with the owner's project requirements that were documented during the planning stage of the project and served as the basis of design.

The goal of project commissioning is to deliver a complete building to the owner, where building equipment and systems operate as required to meet the needs and requirements of the owner and building occupants. This is accomplished not only through design and construction but also through testing and inspection, training of the owner's building operating personnel, and postoccupancy monitoring and testing of building system performance during the warranty period.

9.4 OWNER BUILDING COMMISSIONING BENEFITS

Building commissioning is normally required for third-party certification of a building as a green building. Therefore, some level of building commissioning may be necessary to meet the owner's contract requirement that the building be certified or certifiable as a green building. However, beyond meeting third-party green building certification requirements, building commissioning also provides several real benefits to the building owner. The owner should consider these benefits when deciding whether a building will benefit from commissioning and the extent to which a building should be commissioned. These benefits may result in a more extensive commissioning process than the minimum hurdle required by the third-party green building certification process. In addition, the contractor can use these benefits to encourage the owner to consider a more extensive commissioning process than what the owner might otherwise require. The investment in building commission often makes good business sense for the owner, even though the benefits are often difficult to quantify and the payback time on investment is difficult to predict.

The benefits of building commissioning for the owner include the following:

- Verified equipment and system operation
- Established baseline system performance
- Reduced operating costs
- Improved occupant well-being
- Increased system reliability and maintainability

The following sections discuss each of these advantages of building commissioning.

9.4.1 Verified Equipment and System Operation

First and foremost, a comprehensive commissioning process will verify that each piece of equipment and system meets the owner's operational requirements as defined during the planning stage of the project. Traditionally, at the end of the project, all that is really verified is that equipment has been installed in accordance with the contract requirements and systems are tested as required by the contract documents or local building codes and regulations. This is usually accomplished by an inspection by the owner and design team that is typically focused on installation workmanship, completion of the installation, and basic operation. The results of this inspection are normally summarized in a punchlist, which is a list of items that need to be completed before final project completion and payment.

A traditional end-of-the-project inspection and resulting punchlist does not tie equipment and system performance back to the owner's stated operational requirements. It is usually assumed that the owner's operational requirements are embodied in the design, and if the contractor has procured and installed the required materials and equipment in accordance with the design, then the owner's operational requirements will be met. Commissioning does not make this assumption but instead tests this assumption by using the owner's operational requirements rather than the design as the foundation for the commissioning process. As a result, the ability of the design to fulfill the owner's operational requirements is evaluated during commissioning just like installation. In the end, commissioning will verify that the installed system meets the owner's operational requirements or provide the basis for understanding why the system does not operate as planned.

9.4.2 Established Baseline System Performance

Commissioning will establish a baseline for system and building performance. This baseline or benchmark can be used to evaluate equipment and system performance over the life of the building. Commissioning should not stop at building final completion or even at the end of the warranty period. Commissioning for the owner should be an ongoing and evolving process that ensures that the building continues to meet the owner's current and evolving needs. Building use is seldom stagnant and is always changing with changes in the owner's business or building tenants. Equipment and system performance operation should be monitored throughout the life of the building and adjustments made as required. Similarly, when major changes in building use or occupancy occur, equipment and systems serving the affected areas should be reevaluated to ensure that they are capable of providing the required service and that they are neither too large nor too small. In addition, major changes in space use or new tenants, chronic operational problems, changes in operating philosophy, or other issues may trigger the recommissioning of existing systems, and the original building commissioning would provide the starting point.

9.4.3 Reduced Operating Costs

Building commissioning should result in reduced operating costs for the owner. These reduced operating costs include not only reduced energy costs but also reduced operation and maintenance costs over the life of the facility. Ensuring that equipment and systems are operating properly, as well as that the building is operating as a system, means greater efficiency in operation and less maintenance.

9.4.4 Improved Occupant Well-Being

Commissioning can also result in improved occupant well-being over the life of the building, which in turn will result in greater occupant productivity, fewer building-related illnesses, and higher morale. Buildings establish the physical environment within which occupants live, work, and play, and modern HVAC systems, lighting and lighting control systems, daylighting, and other building environmental systems all contribute to occupant well-being. Commissioning makes sure that all of these systems work together to establish a healthy environment for occupants.

9.4.5 Increased System Reliability and Maintainability

Increased system reliability and maintainability is an outcome that the building owner can expect from building commissioning. First, because equipment was inspected and tested to make sure that it was installed in accordance with the manufacturer's' recommendations, installation issues that may have resulted in reliability and maintenance problems after the building has been occupied may have been resolved. Second, commissioning typically includes training the owner's operating personnel on how to operate and maintain the system. Therefore, operating personnel are more knowledgeable about system operation with commissioning than they would be otherwise. Training also leads to better preventive and predictive maintenance over the life of the building, which translates into reduced downtime and repair costs.

9.5 COMMISSIONING AS A QUALITY ASSURANCE PROGRAM

9.5.1 Contractor Building Commissioning Benefits

Building commissioning will also benefit the contractor. The building commissioning process is really a blueprint for a quality assurance (QA) program that will reduce waste and rework during construction, improve customer satisfaction with the completed building, and reduce callbacks and warranty work after owner occupancy. The commissioning process can serve as the cornerstone of the contractor's QA program, because the objective of commissioning is to ensure that the building operation meets the owner's stated requirements. Commissioning makes good business sense for the contractor whether or not it is required by the contract documents.

9.5.2 What Is Quality?

The term *quality* means many things to many people. The American National Standard ANSI/ISO/ASQ Q9000-2000 (Q9000) entitled *Quality Management Systems – Fundamentals and Vocabulary* defines quality as the "degree to which a set of inherent characteristics fulfills requirements." The term *characteristics* is defined as any "distinguishing feature" and the term *requirements* is defined as any "need or expectation that is stated, generally implied, or obligatory." "Stated needs" refers to the building operational requirements explicitly called out in the contract documents. "Generally implied needs" refers to voluntary industry standards or industry customs and practices to which the owner would normally expect the contractor to adhere. "Obligatory needs" refers to codes, standards, and other documents referenced in the contract documents or required by law.

The Construction Industry Institute (CII) Quality Management Task Force has also similarly defined quality in construction as follows:

> Quality is conformance to established requirements.

This definition agrees with the AGC's definition of quality, which is expressed as simply "conformance to standards." In all cases, the definition of quality is focused on meeting the owner's stated needs, and this is precisely the objective of the building commissioning process.

9.5.3 Quality Control

Quality control (QC) is defined in Q9000 as that "part of quality management focused on fulfilling quality requirements." The terms quality assurance and quality control are often used interchangeably. Quality assurance and quality control are not the same. Quality control is just one part of the contractor's quality assurance program.

Quality control ensures that customer requirements are met through inspection and testing after the building has been completed. The punchlist is a common quality control activity in building construction. The punchlist gives the owner and design team an opportunity to review the completed work and determine if it is in compliance with the contract documents before accepting it. Any identified deviation from the plans and specifications is reviewed, and if warranted, action to correct the deficiency is taken. Quality control leads to very expensive correction of quality problems or disputes, which could be avoided with an effective quality assurance program that includes building commissioning. The punchlist is not to be used as a substitute for an effective QA program.

9.5.4 Quality Assurance

Unlike quality control, which is reactive, quality assurance is proactive. Quality assurance is a broad term that refers to the development and application of procedures that ensure that a product or service meets the customer's stated performance criteria for a building project. *Quality assurance* is defined in Q9000 as that "part of quality management focused on providing confidence that quality requirements will be fulfilled." Quality assurance ensures that the customer's needs and expectations will be met through processes designed to achieve them. Quality assurance is about giving the owner confidence that the contractor can meet his or her needs and expectations per the contract documents.

The owner's performance criteria for a building are defined by the contract documents. The contractor's quality assurance program is concerned with ensuring compliance with the contract documents through the systematic planning, monitoring, and control of the construction process on an ongoing basis. The goal of quality assurance is that the finished work in place complies with the contract documents, avoiding costly and time-consuming rework to correct deficiencies. The goal of quality assurance is very closely aligned with the purpose and objectives of building commissioning.

9.5.5 Quality Planning

Quality planning is defined in Q9000 as "that part of quality management that is focused on setting quality objectives and specifying necessary operational processes and related resources to fulfill the quality objectives." Quality planning is key to the success of the contractor's quality assurance program because it establishes the quality objectives and how they will be achieved. As will be seen later in this chapter, quality planning as defined in Q9000 is synonymous with the commissioning plan, which is the basis of the commissioning process.

9.6 CONTRACTOR'S RESPONSIBILITY FOR BUILDING PERFORMANCE

9.6.1 Importance of Understanding Responsibility for Building Performance

Simply stated, building commissioning is about ensuring that the overall performance of the completed building meets the owner's requirements as documented during the planning stage of the project. Building commissioning involves making sure that the systems that constitute the building can individually and collectively be operated and maintained as required by the owner. Design, procurement, and construction can individually and collectively impact equipment, system, and building performance. Deficiencies found during the building commissioning process must be resolved, which means that the reason for the deficiency needs to be ascertained, corrective action taken, and the component or system retested to demonstrate compliance. Taking action to correct a deficiency and then retesting can result in additional costs for the contractor, as well as impact productivity or delay project completion. Therefore, the contractor needs to have a clear understanding of its responsibilities for equipment, system, and building performance so that any additional time or cost incurred correcting deficiencies during the commissioning process can be recovered through a change order if the deficiency was not the contractor's responsibility.

9.6.2 Determining Responsibility for Building Performance

Responsibility for building performance will be a matter of contract between the owner and contractor. To determine responsibility for the performance of a piece of equipment, system, or overall building, the contractor needs to first look at its owner-contractor agreement, including all documents referenced. In most cases, the project specifications will determine responsibility for equipment, system, or building performance on design-bid-build projects, and the owner's project criteria will determine performance requirements on design-build projects. If there is a conflict between contract documents regarding performance, the contractor should look to the order of precedence included in the owner-contractor agreement (as discussed in Chapter 5). Similarly, mixed specifications are also covered in Chapter 5, where materials and installation are specified as well as system performance. Although project-specific contract provisions will govern and need to be reviewed to determine equipment, system, and building performance on a particular building project, the type of project delivery system employed on a project will provide some guidance regarding performance responsibility. The following sections briefly discuss performance responsibility in general for these project delivery scenarios:

- Design-Bid-Build: Descriptive or Prescriptive Specification
- Design-Bid-Build: Performance Specification
- Design-Build
- Draw-Build or Design-Assist

Design-Bid-Build: Descriptive or Prescriptive Specification. Under the traditional design-bid-build approach, the design team designs the system and specifies the materials and equipment using either a descriptive or prescriptive specification (as discussed in Chapter 5). Under this scenario, the owner assumes the design risk for equipment, system, and building performance through the owner-contractor agreement. As long as the contractor operating as either a general contractor or construction manager at-risk procures and installs the materials and equipment in accordance with the contract documents, the owner retains the risk of system performance. The contractor is only responsible for achieving the owner's operational requirements as set forth in the contract documents including the plans and specifications. As a result, if during the commissioning process the equipment or system performance is found to be deficient, the contractor should not be responsible for additional costs and time required to resolve the deficiency.

Design-Bid-Build: Performance Specification. Using a performance specification with a traditional design-bid-build approach, the contractor does assume the design risk for system performance along with the installation risk. The contractor designs, procures, and installs the system based on the performance specifications included in the contract documents. However, the contractor is only responsible for achieving the owner's operational requirements as set forth in the performance specification by the designer. Any discrepancy between the owner's operational requirements and the designer's performance specification will be the owner's responsibility. As a result, if during the commissioning process the equipment or system performance is found to be deficient because of performance specification requirements, the contractor should not be responsible for additional costs and time required to resolve the deficiency. However, if the performance deficiency discovered is the result of the contractor's installation not meeting the performance specification, then it will be the contractor's responsibility to resolve the deficiency.

Design-Build. If the project delivery system employed is design-build, then the contractor operating as the design-builder will be responsible for overall building and system performance in accordance with the owner's documented performance criteria. The owner's performance criteria will usually be included as part of its request for proposal (RFP), which normally becomes part of the project contract documents on design-build projects. On a design-build project, the contractor will be responsible for correcting any equipment, system, or building performance deficiencies identified during the commissioning process based on the owner's documented design criteria. One exception to this would be a design-build project where the contractor's design documents that are approved by the owner take precedence over the owner's design criteria in the owner–design builder agreement.

Draw-Build or Design-Assist. Where responsibility for equipment, system, and building performance is fairly straightforward on a pure design-build project as described in the previous section, responsibility may become murky where design-build variations like draw-build or design-assist are employed. Draw-build or design-assist usually refer to the scenario where the owner provides the contractor with construction documents, including both drawings and specifications, that are typically 30 to 35 percent complete through the schematic design phase. It is the contractor's responsibility to complete the design and construct the facility based on the completed design.

With these design-build variations, there may be confusion as to where the owner's responsibility for system performance leaves off and the contractor's responsibility for system performance begins. The contractor needs to make sure it fully understands its contractual responsibilities for the equipment, system, and building performance under either draw-build or design-assist. In other words, is the contractor assuming responsibility for ensuring that the owner's design will meet the performance requirements? Or is the contractor only assuming responsibility for detailing the remaining portions of the system and that it works with the owner-specified systems and equipment? Responsibility for performance deficiencies encountered during commissioning a draw-build or design-assist project may be more difficult to sort out than with any of the other scenarios, and the contractor needs to be aware of this situation.

9.7 BUILDING COMMISSIONING EXTENT AND TIMING

The extent and timing of building commissioning depends on the project. On a traditional building project, systems are checked out and started up by the installing contractor during construction. The installing contractor usually retains documentation unless specifically required to be submitted to the owner by the contract documents. In addition, each system is usually checked out independently, and there is often no cohesive plan for inspection, testing, and startup to verify that the various systems will operate together, except in the case of life-safety systems as dictated by local codes and the local authority having jurisdiction. In the end, the owner and design team inspect the project and prepare a punchlist of things that need to be completed before final project closeout.

On a green building project, building commissioning is typically more extensive and involves not only the installing contractor but also an entire commissioning team on some projects. Commissioning often includes a commissioning authority, which is usually an outside individual or firm retained directly by the owner, whose responsibility it is to develop and execute the commissioning plan with the help of the design and construction teams.

Commissioning can start anytime during the project. Commissioning can take place as late as at the end of construction. However, waiting until the end of the project may be too late to make needed changes in an economical and timely manner. Ideally, commissioning should start with design intent and extend through the warranty period with actual verification of system performance.

9.8 PROJECT COMMISSIONING REQUIREMENTS

Project commissioning requirements will usually be included in the project specifications and may also be included in a separate draft commissioning plan prepared by the owner's commissioning authority during the design phase of the project. If there is a separate draft commissioning plan, it should be coordinated with the project specifications. The draft commissioning plan should be included in the owner's request for bid or proposal documents, so that the contractor will understand the extent of the commissioning process and its involvement, and this can be incorporated into its bid or proposal to the owner. In addition, the draft commissioning plan should be included in the owner-contractor agreement as a reference document. The inclusion of the draft commissioning plan in the owner-contractor agreement can be accomplished by specifically naming it as a contract document in the contract or by reference in the project specifications.

9.8.1 Contractor Needs to Understand Commissioning Requirements

The contractor needs to thoroughly understand the scope of the building commissioning requirements as well as its role and responsibility for the commissioning process before submitting a bid or proposal for the project. Commissioning can be a very labor-intensive and time-consuming process that requires not only resources and time to perform the work but also significant project management and field supervision time as well as clerical and drafting support. Commissioning has the potential of impacting direct construction costs and the project schedule as well as the contractor's project overhead.

9.8.2 Defining Subcontractor Scopes of Work

The contractor also needs to thoroughly understand the project commissioning requirements so that it can include commissioning requirements in subcontractor scopes of work to ensure a complete and accurate subcontract bid or proposal that includes the commissioning activities, commissioning documentation, and other commissioning requirements for which the subcontractor will be responsible. If the contractor fails to understand the project commissioning requirements and does not adequately address them when defining subcontract scopes of work, it may find itself responsible to the owner for the work through its owner-contractor agreement and required to issue a change order to one or more of its subcontractors for the required commissioning work that was not included in their subcontract scope of work. The contractor should also ensure that it has included subcontractor commissioning requirements in its subcontracts and that there are no gaps or overlaps between its commissioning scope of work and that of its subcontractors or between the subcontractors themselves.

9.8.3 1995 CSI MasterFormat™

The 16-division 1995 edition of the Construction Specifications Institute (CSI) MasterFormat™ is still used extensively in the building construction industry, even though a new edition was published in 2004 by CSI. A listing of the 16 divisions included in the 1995 CSI MasterFormat™ is provided in Table 9-1. Because building commissioning was not common in the early to mid-nineties, the 1995 CSI MasterFormat™ does not address commissioning to the extent that the 2004 edition does.

Table 9-2 provides a listing of sections in the 1995 CSI MasterFormat™ where building commissioning requirements might be found. However, the CSI MasterFormat™ is only a suggested outline for project specifications, and the specification drafter is not bound by its use. Therefore, commissioning requirements may be found elsewhere in the project specifications or in other contract documents.

In addition to the 16 divisions contained in the 1995 edition of the CSI MasterFormat™, it is common for specification writers to include a separate Division 17 for systems and activities not covered in the other divisions. Among other uses, a separate Division 17 is sometimes used to specify requirements for building commissioning that may not fit elsewhere in the 1995 CSI MasterFormat™ or to group all general commissioning

Table 9-1

CSI 1995 Masterformat™ List of Divisions

Division 1	General Requirements
Division 2	Site Construction
Division 3	Concrete
Division 4	Masonry
Division 5	Metals
Division 6	Wood and Plastics
Division 7	Thermal and Moisture Protection
Division 8	Doors and Windows
Division 9	Finishes
Division 10	Specialties
Division 11	Equipment
Division 12	Furnishings
Division 13	Special Construction
Division 14	Conveying Systems
Division 15	Mechanical
Division 16	Electrical

Table 9-2

CSI 1995 Masterformat™ Commissioning Requirements	
DIVISION 1–GENERAL REQUIREMENTS	
01700	Execution Requirements
01750	Starting and Adjusting
01770	Closeout Procedures
01780	Closeout Submittals
01800	Facility Operation
01810	Commissioning
01820	Demonstration and Training
01830	Operation and Maintenance
DIVISION 15 – MECHANICAL	
15900	Automatic Controls
15950	Testing, Adjusting, and Balancing
DIVISION 16 – ELECTRICAL	
16000	General
16080	Electrical Testing

requirements in one location for easy reference and use. Including a separate division outside of the CSI-prescribed 16 divisions can help the contractor locate and understand building commissioning requirements.

9.8.4 2004 CSI MasterFormat™

As noted, the 2004 edition of the CSI MasterFormat™ does explicitly address building commissioning throughout the building construction divisions. The 2004 CSI MasterFormat™ building construction divisions organized by subgroup are listed in Table 9-3. As can be seen from Table 9-3, there are now 21 building construction divisions. The first 14 are basically the same as the previous 1995 CSI MasterFormat™, and the remaining 7 divisions make up the Facility Service Subgroup. The 2004 CSI MasterFormat™ Facility Service Subgroup replaces the former Division 15/Mechanical and Division 16/Electrical as well as part of the former Division 13/Special Construction that addressed security access and surveillance, building automation and control, fire alarm, and fire suppression systems.

General commissioning requirements can be found in Division 01/General of the 2004 CSI MasterFormat™. Section 01 90 00 addresses building life-cycle activities and includes the Level 3 and Level 4 requirements shown in Table 9-4. As can be seen from Table 9-4, Section 01 91 13 addresses general commissioning requirements.

Table 9-3

CSI 2004 Masterformat™ List of Building Divisions by Subgroup
GENERAL REQUIREMENTS SUBGROUP
Division 01 – General Requirements
FACILITY CONSTRUCTION SUBGROUP
Division 02 – Existing Conditions
Division 03 – Concrete
Division 04 – Masonry
Division 05 – Metals
Division 06 – Wood
Division 07 – Thermal and Moisture Protection
Division 08 – Openings
Division 09 – Finishes
Division 10 – Specialties
Division 11 – Equipment
Division 12 – Furnishings
Division 13 – Special Construction
Division 14 – Conveying Equipment
FACILITY SERVICE SUBGROUP
Division 21 – Fire Suppression
Division 22 – Plumbing
Division 23 – Heating, Ventilating, and Air-Conditioning (HVAC)
Division 25 – Integrated Automation
Division 26 – Electrical
Division 27 – Communications
Division 28 – Electronic Safety and Security

Product- and system-specific commissioning requirements are addressed in their respective divisions as shown in Table 9-5. In the 2004 CSI MasterFormat™, the Level 3 "08" section of each division is dedicated to commissioning requirements. As building commissioning becomes more common and spreads to other than the building mechanical, electrical, and plumbing (MEP) systems, the commissioning sections in the Facility Construction Subgroup will become more important.

9.8.5 Division 25—Integrated Automation

Division 25 entitled Integrated Automation (IA) was added as a new division in the Building Services Subgroup by CSI in the 2004 edition of the MasterFormat™. While Division 25 is new and not yet used extensively, it represents a growing trend in the building industry toward system integration

Table 9-4

CSI 2004 Masterformat™ 01 90 00 Life-Cycle Activities General Commissioning Requirements	
01 91 00	Commissioning
01 91 13	General Commissioning Requirements
01 92 00	Facility Operation
01 92 13	Facility Operation Procedures
01 93 00	Facility Maintenance
01 93 13	Facility Maintenance Procedures
01 93 13	Recycling Programs
01 94 00	Facility Decommissioning
01 94 13	Facility Decommissioning Procedures

Table 9-5

CSI 2004 Masterformat™ Specific Commissioning Requirements	
02 08 00	Existing Conditions
03 08 00	Concrete
04 08 00	Masonry
05 08 00	Metals
06 08 00	Wood, Plastics, and Composites
07 08 00	Thermal and Moisture Protection
08 08 00	Openings
09 08 00	Finishes
10 08 00	Specialties
11 08 00	Equipment
12 08 00	Furnishings
13 08 00	Special Construction
14 08 00	Conveying Equipment
21 08 00	Fire Suppression
22 08 00	Plumbing
23 08 00	HVAC
25 08 00	Integrated Automation
26 08 00	Electrical
27 08 00	Communications
28 08 00	Electronic Safety and Security

and interoperability as well as open-architecture control systems. The goal of IA is to optimize the overall operation of the building in order to provide a healthier and more productive environment for occupants as well as to increase the efficiency of building operations. The realization of this goal requires the integration of all building systems under the control of the IA system so that building operation can be optimized as a whole. IA will impact the commissioning process, and the contractor needs to be aware of what Division 25 covers.

Division 25 specifies the IA system that integrates all of the subsystems represented by the Facilities Subgroup, which includes Division 21/Fire Suppression, Division 22/Plumbing, Division 23/HVAC, Division 26/Electrical, Division 27/Communications, and Division 28/Electronic Safety and Security, along with Division 11/Equipment and Division 14/Conveying Systems. Division 11 covers equipment that serves a unique function in a building such as foodservice, laboratory, or athletic equipment. Conveying systems such as elevators and escalators are covered in Division 14.

All of the hardware and software needed to implement an IA system are specified in CSI Division 25. This includes conductors and raceways, network equipment such as servers and hubs, instrumentation and terminal devices that interface directly with building equipment or through system-specific devices specified elsewhere, gateways to establish a communications link between the IA system and other stand-alone building systems, and control sequences that describe how the IA system is to operate.

IA will cover future building automation and control systems whether or not they remain proprietary or shift to open-architecture systems. An open-architecture control system is one where the hardware and software specifications are public information and available to anyone who wants to manufacture hardware components or develop software for the system. This is in contrast to closed-architecture or proprietary control systems, where the original system developer maintains control of the system specifications and is the only entity that can supply hardware or software for the system. Today, proprietary building automation, fire alarm, security, and other systems are the norm, and the manufacturers of these systems not only supply the system but also perform the original system installation and provide ongoing system maintenance.

The use of integrated open-architecture building automation and control systems will increase the complexity of the building commissioning process. For one, the contractor will probably have a separate contract with a systems integration firm that will design, install, and check out the system. Where traditional proprietary control systems include standard algorithms and control sequences, open-architecture control systems are mainly customized for the building in which they are installed. As a result, functional testing, which includes not only testing the operation of the system being commissioned

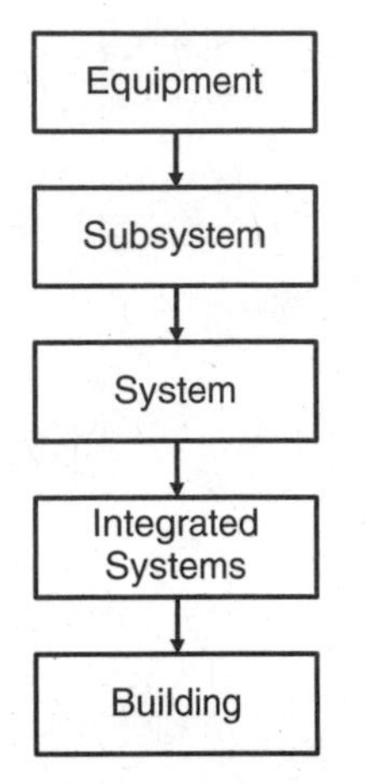

Figure 9-1 Basic Building Commissioning Process.

but also its interaction with other related building systems, will probably be more complex and detailed.

9.9 BASIC BUILDING COMMISSIONING PROCESS

Building commissioning is about inspecting, testing, and documenting the performance of individual building equipment, systems, and system integration. Today, commissioning focuses primarily on the mechanical, electrical, life-safety, and security systems in buildings. However, commissioning can also apply to other building systems, such as the building envelope, and probably will in the future as these systems gain intelligence and are integrated into the building control system. The building commissioning hierarchy is illustrated in Figure 9-1.

9.9.1 Equipment Commissioning

A commissioning program starts with ensuring that individual pieces of equipment have been installed, inspected, prepared for energization or startup, and energized or started up in accordance with the contract documents, applicable codes and standards, and manufacturers' recommendations. Testing may also be performed to determine if the equipment is operating in accordance with contract requirements and manufacturer guarantees. Usually, each piece of equipment that is part of a system that will be commissioned is inspected and tested in accordance with a written procedure and the results documented for verification and for future reference.

9.9.2 Subsystem Commissioning

The complexity of building systems is increasing, and once stand-alone systems are being integrated as subsystems into more comprehensive building systems. For instance, in the past the building fire alarm system was often a completely separate system with no relationship to the building security system. In most modern buildings today, the security and fire alarm systems are viewed as interdependent systems with the common mission of protecting building occupants and preventing property loss. Even though each of these systems have its own unique functions, they are both part of the overall building life-safety system. From a commissioning standpoint, the operation of individual fire alarm functions such as alarm annunciation and security system functions such as intrusion detection should be verified before verifying the interaction between these systems and the overall function of the integrated building life-safety system.

9.9.3 System Commissioning

Building systems comprise individual pieces of equipment and subsystems that must work together to ensure that the owner's operational requirements are met. Even though an individual piece of equipment or subsystem is found to operate properly, there is no guarantee that the overall building system will operate optimally or even acceptably. The next step in the commissioning process is to ensure that building systems operate as planned. This involves testing overall building systems to ensure that individual pieces and equipment and subsystems can work together to achieve the overall building system performance objectives.

For example, a building that includes telecommunications equipment that needs to continue operation in the event of a power outage resulting from a utility brownout or blackout may include both an emergency generator and an uninterruptible power supply (UPS) system. Upon loss of power, the emergency generator's transfer switch is required to detect the loss of power, start up and synchronize the emergency generator, and then transfer the critical building loads such as the telecommunications equipment from the utility power supply to the emergency generator. Similarly, upon loss of power, the UPS system would supply the telecommunications equipment from its batteries. Both of these subsystems would be tested independently to ensure their operation. However, as part of the commissioning process, these two subsystems, as well as other related subsystems, should be tested together to make sure the entire building emergency power system operates as required. This could be accomplished by opening the building service disconnect to test the building emergency power system operation.

9.9.4 Integrated Systems Commissioning

The next step in the building commissioning process should be ensuring that building systems are integrated and work together as needed. An example of integrated building systems commissioning would be testing to make sure that the building life-safety and HVAC systems function properly upon detection of smoke in a return air duct by a duct smoke detector. Detection of smoke in a return air duct should not only trigger the fire alarm but also result in a defined response from the HVAC system to isolate the affected zone through the operation of smoke dampers and turning on fans and opening dampers to exhaust smoke from the affected zone. Similarly, the building life-safety system will also be integrated with the elevator system, and the detection of smoke may result in elevators bypassing the affected floor and returning to the main floor and locking out until firefighters arrive.

9.9.5 Building Commissioning

The ultimate goal of the commissioning process is to ensure that the entire building operates efficiently as a system. In the past, building systems were designed, installed, and operated as independent systems. While the performance of individual building systems may be optimized, when combined with other related building systems, the overall building operation may be suboptimal. Green or high-performance buildings recognize this situation, and their goal is to optimize the operation of the building as a whole instead of individual building systems. Determining if the overall building is operating optimally is usually accomplished after the building is occupied by monitoring building operations using the data acquisition and analysis capabilities of the building management system (BMS) or by installing stand-alone data loggers to measure specific building parameters.

9.10 COMMISSIONING AUTHORITY

9.10.1 What Is the Commissioning Authority's Role?

The commissioning authority (CA or CxA) is the individual or organization that leads the building commissioning effort. The commissioning authority is responsible for developing the building commissioning process and documenting it as the commissioning plan and then ensuring that the building commissioning is carried out in accordance with the plan.

9.10.2 Who Can Be the Commissioning Authority?

The commissioning authority can be any qualified individual or firm retained directly by the owner or through the architect or contractor. Normally, the commissioning authority will be an independent third party that contracts directly with the owner to plan, coordinate, implement, and document the commissioning process. In general, third-party green building certification criteria requires that the commissioning authority be an independent third party with no employment or contractual relationship with the design team or the construction team. The exception to this is small buildings, where systems are relatively simple and building commissioning is straightforward.

Because the commissioning authority is charged with developing and overseeing the commissioning process for the owner, which involves verifying that the installed equipment, systems, and building performance meet the owner's stated performance criteria, it is best that the commissioning authority be an independent third party with no ties to either the design or construction teams. Owners and designers usually understand the need for the commissioning authority to be independent of the construction team, because the commissioning process involves evaluating the equipment and systems installed by the construction team.

Deficiencies discovered during the commissioning process can be expensive to correct and result in delays in project completion. Having a commissioning authority that is employed by or contracted to the contractor or subcontractor could result in a real or perceived conflict of interest. There could also be concern about bias if the design is deemed to be partially or wholly responsible for the performance deficiency. Also, the commissioning authority should communicate with and report directly to the owner, which would also be difficult if the commissioning authority was employed by or contracted to the construction team.

It is not always so readily apparent to owners why the commissioning authority should also be independent of the design team. This stems from the fact that, traditionally, the architect and other members of the design team have assumed the role of the owner's watchdog during construction, making sure that the contractor's work is performed in accordance with the contract documents. The commissioning authority's role in a green building project is different and goes beyond the design team's traditional role of watchdog. The commissioning authority is responsible for ensuring that the completed building performs in accordance with the owner's operational requirements and not just that the equipment and systems were installed and tested in accordance with the plans and specifications. As a result, the commissioning process evaluates the adequacy of the design team's design as well as ensures that the construction team performed the installation correctly.

For this reason, the commissioning authority should not be employed by or contracted to any member of the design team or the construction team.

9.10.3 Contractor as Commissioning Authority

As noted, the contractor can act as the commissioning authority for the owner if it is not barred by the criteria being used for third-party green building certification. It is unlikely that the contractor will perform this work itself, because few general contractors or construction managers would have the in-house expertise to act as commissioning authority. Instead, the contractor will most likely subcontract the role of commissioning authority to either an individual or firm with the necessary resources and expertise to perform the work.

The contractor does include the role of commissioning authority in the scope of work to one of its first-tier subcontractors such as the mechanical contractor. If this is done, it should be done with the stipulation that the first-tier subcontractor cannot subcontract the work to another contractor that would be a second-tier subcontractor to the contractor. Conceivably, the first-tier mechanical subcontractor could subcontract the work to a second-tier sheet metal contractor, who in turn subcontracts the work to a third-tier testing, adjusting, and balancing (TAB) contractor.

Also, the contractor should avoid contracting directly with a second- or third-tier subcontractor for the commissioning authority work. Any of these arrangements would be very difficult administratively, might result in a perceived or actual conflict of interest, and would make it nearly impossible for the commissioning authority to carry out its work. The contractor should avoid subcontracting the commissioning work to any of its construction subcontractors if possible, and certainly not allow it to be performed by other than a first-tier contractor if an outside commissioning authority cannot be found.

9.10.4 Commissioning Authority Qualifications

The commissioning authority should be an individual or firm that is experienced in the planning, design, installation, startup, and operation of mechanical, electrical, and plumbing (MEP) systems as well as building automation and control systems. In addition, the commissioning authority needs to fully understand its role in the commissioning process and preferably have previous experience in building commissioning. If the owner is seeking third-party certification of the building as a green building, then the commissioning authority definitely needs a thorough understanding of the criteria and rating system that will be used for building certification. As can be seen from the previous description of needed commissioning authority qualifications, it would be very difficult for an individual to be a commissioning authority on a building of any size. Most of the time, the commissioning authority is a firm that specializes in building commissioning, an MEP design firm that offers commissioning as a service to clients, or a specialty contractor that has in-house design and construction expertise.

9.10.5 Commissioning Authority Responsibilities

Table 9-6 provides a list of typical commissioning authority responsibilities. Many of these typical responsibilities are discussed in the sections that follow. As can be seen from the list, the commissioning authority's responsibilities are

Table 9-6

Typical Commissioning Authority Responsibilities
• Plan and coordinate the commissioning process.
• Develop a draft commissioning plan.
• Submit the draft commissioning plan for review by the owner and designer.
• Revise the commissioning plan.
• Review the commissioning plan with the contractor and affected subcontractors.
• Finalize and distribute the commissioning plan.
• Establish and coordinate the commissioning team.
• Collect equipment and system information from subcontractors.
• Develop prefunctional inspection, testing, and startup (PITS) procedures.
• Plan and coordinate PITS.
• Witness PITS performance and conduct spot checks as required.
• Review and approve PITS outcomes and documentation.
• Plan and coordinate control system testing.
• Review and approve control system testing outcomes and documentation.
• Plan and coordinate HVAC testing, adjusting, and balancing (TAB).
• Review and approve HVAC TAB outcomes and documentation.
• Develop detailed functional system testing.
• Plan and coordinate functional system testing.
• Review and approve functional testing outcomes and documentation.
• Review and approve operations and maintenance (O&M) documentation.
• Plan and coordinate owner personnel training by contractor and subcontractors.
• Determine building functional completion.
• Perform other responsibilities as required.

extensive and should start during the design stage of the building project. This means that if the owner plans to have the contractor act as commissioning authority, the owner should use a construction management approach to project delivery that involves the contractor throughout the design process rather than the general contractor approach where the contractor is brought in after the design is completed.

9.11 COMMISSIONING PLAN

The commissioning plan is central to the commissioning process. The commissioning plan is typically prepared by the commissioning authority during design development and addresses commissioning activities during the design, construction, acceptance, and warranty phases of the project. The initial commissioning plan evolves throughout design and construction as more detailed information about the equipment and systems to be commissioned becomes available. The initial commissioning plan prepared by the commissioning authority is also used by the design team to integrate the general commissioning requirements in the initial commissioning plan into the project specifications and other contract documents. Figure 9-2 summarizes the typical commissioning plan development process.

9.12 COMMISSIONING TEAM

The commissioning team consists of representatives from the owner, design, and construction teams. The exact makeup of the commissioning team will depend on the project, project delivery system, extent of commissioning required, and commissioning process. The commissioning team is responsible for implementing the commissioning plan. For a typical building project, the commissioning team will usually consist of the following:

- Commissioning Authority
- Owner's Representative
- Contractor
- Mechanical Contractor
- Electrical Contractor
- Testing, Adjusting, and Balancing (TAB) Contractor

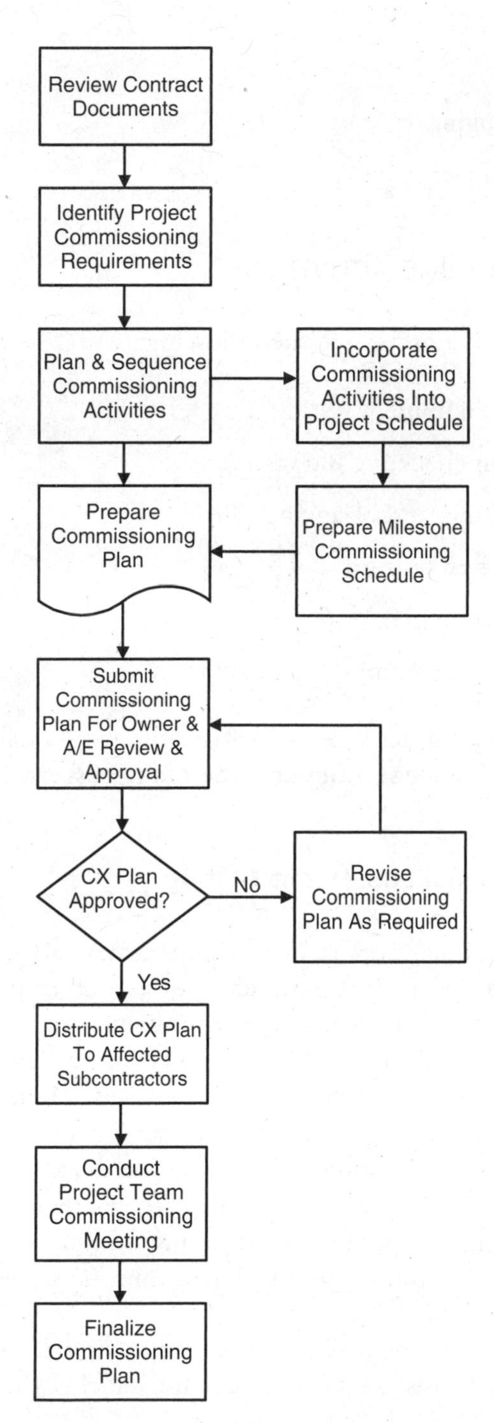

Figure 9-2 Commissioning Plan Development.

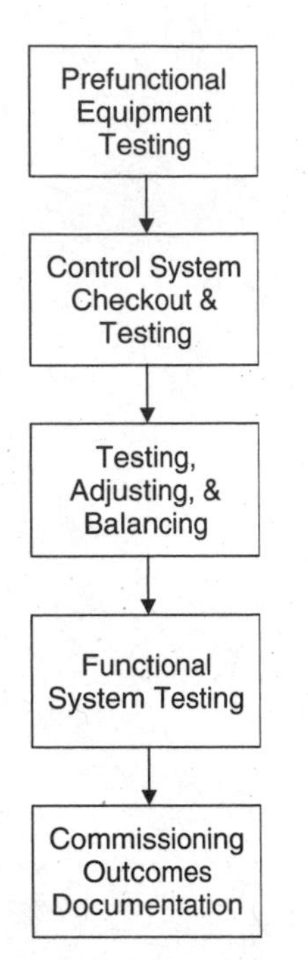

Figure 9-3 Commissioning Activities.

- Controls Contractor
- Architect
- Mechanical Engineer
- Electrical Engineer

9.13 COMMISSIONING ACTIVITIES

Commissioning activities include the following:

- Prefunctional equipment testing
- Control system checkout and testing
- Testing, adjusting, and balancing (TAB)
- Functional system testing
- Operational training
- Commissioning outcomes documentation

These commissioning activities are carried out sequentially as illustrated in Figure 9-3. The following sections describe and discuss each of these activities.

9.13.1 Prefunctional Equipment Testing

What Is Prefunctional Equipment Testing? The purpose of prefunctional equipment testing is to ensure that individual pieces of equipment operate as planned. Prefunctional equipment testing involves inspection, testing, and startup of equipment and is usually performed by the installing contractor. Normally, every piece of equipment associated with a building system that is part of the commissioning plan and will undergo functional testing is required to undergo prefunctional testing. Sampling or other statistical techniques for selecting a representative set of pieces of equipment are usually not allowed in prefunctional testing. All prefunctional testing of equipment that comprises a building system that will be commissioned must be successfully completed before functional testing can start.

Who Develops the Prefunctional Testing Requirements? The prefunctional testing plan usually consists of a test procedure and checklist for each piece of equipment. These procedures and checklists incorporate the equipment manufacturer's inspection, testing, and startup requirements as well as project-specific requirements, such as specific equipment operating characteristics that are important for functional system testing. The prefunctional testing plan can be developed by the commissioning authority with input from the installing contractor or by the installing contractor within the guidelines established by the commissioning authority. The process for developing a prefunctional equipment testing plan by a subcontractor is illustrated in Figure 9-4.

How Is Prefunctional Equipment Testing Performed? The prefunctional testing process involves the performance of the following three activities for each piece of equipment:

- Inspection
- Testing
- Startup

Inspection. The equipment installation needs to be inspected by the installing contractor to ensure that the equipment was installed in accordance with the manufacturer's requirements and the plans and specifications. This inspection should not only verify that the physical installation and necessary equipment connections have been made but also make sure that the shipping materials, bolts, and blocking have been removed from mechanical and electrical equipment. For motor-driven equipment, inspection may also include verifying fluid levels, belt tension, gages, and sensor calibration, among other things.

Testing. This testing is typically static testing to make sure that the equipment is ready for startup. For example, many three-phase motors are rated for both 240-volt and 480-volt operation, and the motor's operating voltage will be determined by the connections made by the electrician when the motor is wired. During prefunctional inspection, it should be verified that a fan motor is connected properly for the voltage serving it. During testing, the actual voltage should be verified by measurement at the motor starter before energizing the motor. In addition, phase rotation should also be checked to make sure that the motor's shaft will rotate in the correct direction when the fan is started up. This can be accomplished using a phase rotation meter or by "bumping" the motor. Bumping the motor involves turning it on and off very quickly to verify that the shaft is rotating in the correct direction for the fan.

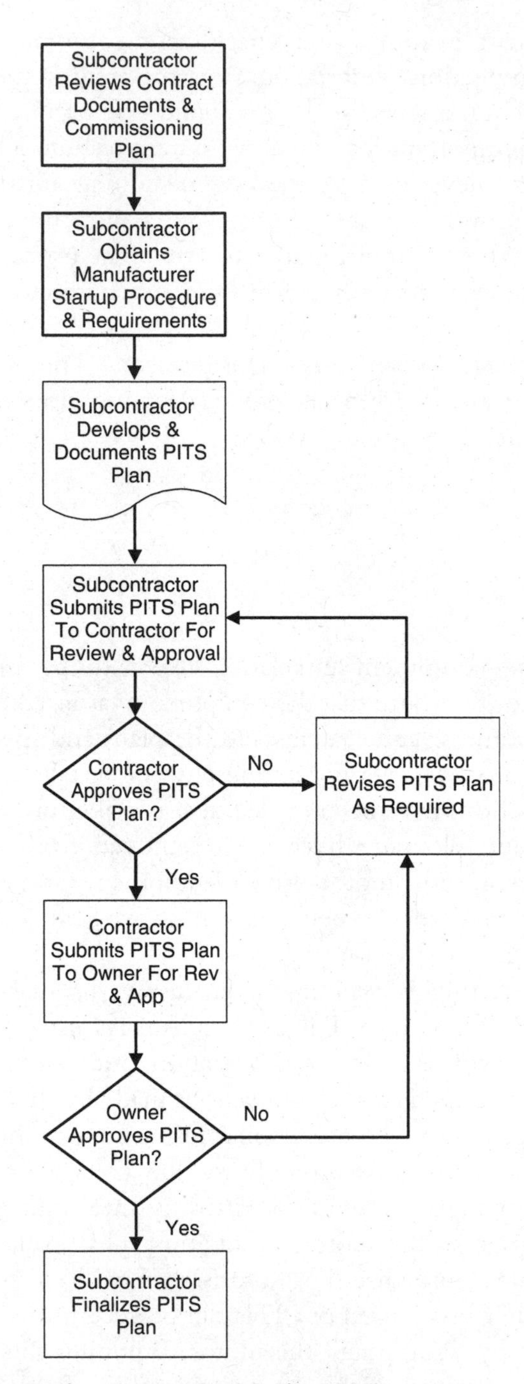

Figure 9-4 Prefunctional Equipment Testing Plan Development.

Startup. Startup or energization is the last step in prefunctional testing. The equipment is started, and once it is running, it is observed and tested to ensure that it is operating properly. In the case of a fan, this testing might include verifying the volume of air moving or differential pressure as well as measuring the current draw of the motor to determine if the fan is operating as specified.

Prefunctional Equipment Testing Documentation. The results of prefunctional testing need to be documented using the checklists developed and submitted to the commissioning authority for review and approval. Any problems encountered during prefuntional equipment testing need to be documented and corrected if they are installation problems or documented for correction later if they are design or manufacturing problems. In addition, the commissioning authority may also want to witness selected prefunctional equipment testing as it is being performed or go back and spot-check equipment testing after the installing contractor has completed the prefuntional equipment testing.

9.13.2 Control System Checkout and Testing

Before performing the HVAC system testing, adjusting, and balancing (TAB) work, the HVAC system must be complete and the HVAC control system must be operational, tested, and documented by the subcontractor performing the HVAC control system work for the building. The HVAC control system subcontractor needs to provide the TAB contractor with the information that it needs to perform its work. This information includes the HVAC system sequence of operation, temperature and pressure setpoints, monitoring and control points, among other information. Typically, the commissioning authority will set up a meeting with the contractor, HVAC controls subcontractor, and TAB subcontractor when the control system installation is nearing substantial completion to identify the information that the TAB subcontractor needs from the HVAC controls subcontractor and begin to plan and schedule the TAB work.

9.13.3 Testing, Adjusting, and Balancing

The purpose of testing, adjusting, and balancing (TAB) the HVAC system is to ensure that the system is operating as designed to provide occupant comfort and energy efficiency. TAB involves measuring and balancing air and hydronic flow rates to bring them into compliance with the design, and then verifying and documenting that these flow rates do conform with the design. This is accomplished by the TAB contractor by adjusting valves and dampers as well as varying fan speeds to achieve the design temperature and air flow under test conditions. The HVAC air and hydronic TAB work must be successfully completed and documented before the start of functional testing.

9.13.4 Functional Testing

What Is Functional Testing? As noted previously, functional testing involves testing of systems, whereas prefunctional testing ensures that the equipment constituting the systems is operating properly. With functional testing, the operation of building systems as well as the interaction of building systems are verified. For example, prefunctional testing will verify that a return air duct smoke detector operates properly upon detection of smoke. During functional testing, the reaction of the fire alarm system to the individual duct smoke detector is tested, along with the fire alarm system's interaction with the HVAC system, elevator system, and other building systems. Functional testing might require that the smoke detector be activated by simulated smoke in the duct and then observe the chain of events that follows to ensure that the fire alarm system reacts as planned with alarms and notifications, as well as interacts properly with other associated building systems. The interaction with the HVAC system could include closing of smoke doors and dampers in the affected area, shutdown of air handling units serving the affected area, opening of smoke exhaust dampers and the startup of smoke exhaust fans, and startup of stairwell pressurization fans, among other things. Additionally, the elevator system might react by recalling all elevator cars to an evacuation floor and locking them out.

Methods of Functional Testing. Functional testing can be performed in either of the following two ways:

- Manual testing
- System monitoring

Manual Testing. Manual testing involves purposely changing a system input or condition and then observing the reaction of the system under test to the changed condition as well as its interaction with other related systems. The previous section provided an example of a manual test performed on a building fire alarm system. In this case, the artificial smoke into a return air duct resulted in the duct smoke detector detecting the smoke and setting off a chain of events through the fire alarm system, which included sounding alarms, notifying the authorities, and causing reactions in both the HVAC and elevator systems.

System Monitoring. System monitoring is passive and involves monitoring the operation of a building system over time to determine how it reacts to changing conditions. System monitoring is often performed using the data acquisition and analysis capabilities of the building management system (BMS) when the needed system monitoring points are available. When monitoring cannot be performed using the BMS, portable data loggers can be used. A portable data logger is a device that can measure and record a variety of system characteristics, such as temperature, humidity, air or water flow, air or water pressure, voltage, current, power, energy, and just about any other measurable quantity by using the proper transducer. An example of system monitoring might be the measurement of the carbon dioxide (CO_2) level in a space to determine if the HVAC system will detect the elevated CO_2 level and increase the amount of outside air in the supply air stream to lower the CO_2 level in the space. Similarly, system monitoring can be used to monitor the reaction of the HVAC system to changes in outside air temperature or space occupancy, the operation of various equipment under normal conditions, among other things.

9.13.5 Operational Training

The commissioning process requires that the owner's operating personnel be trained in the proper and safe operation of the commissioned systems as well as system maintenance and troubleshooting. The contractor will normally meet this commissioning requirement through its subcontractors that installed the system. Normally, this includes the mechanical subcontractor responsible for the HVAC and HVAC control systems, the electrical subcontractor responsible for the power, communications, and lighting control systems, the building automation and control system subcontractor, the life-safety and security system subcontractor, and other commissioned building system subcontractors.

Training the owner's operational personnel will involve planning and scheduling the training, assembling training materials, presenting the training, which can include both classroom instruction and hands-on training, and documenting attendance and training outcomes. The training provided must address the unique features of the building, consider the background and experience of the owner's operating personnel, and be based on the contents of the O&M manuals provided to the owner. The goal of this training is to ensure that the owner's operating personnel not only know how to operate the building systems but also understand how they should operate, what regular routine maintenance and testing needs to be performed, and how to troubleshoot the system when it is not operating properly.

9.13.6 Commissioning and Closeout Submittals

The final step in the commissioning process is for the contractor to gather the required commissioning documentation and submit it in the prescribed format to the owner. This documentation will normally be submitted in the form of O&M manuals, as discussed earlier in this chapter. The contractor needs to understand what information is required and the format it is to be submitted in. The submittal requirements may be found in the project specifications, in the commissioning plan, or both. As shown in Table 9-7, submittal requirements are provided in Section 01 78 00 of the 2004 CSI MasterFormat™. Submittal requirements may also be found in other equipment- or system-specific sections, such as Section 23 05 93, which covers HVC system TAB requirements.

9.14 PROJECT FUNCTIONAL COMPLETION

Project functional completion is a milestone that is unique to commissioned buildings. Project functional completion is usually achieved when the

Table 9-7

CSI 2004 Masterformat™ 01 78 00 Closeout Submittals	
01 78 13	Completion and Correction List
01 78 19	Maintenance Contracts
01 78 23	Operation and Maintenance Data
01 78 23.13	Operation Data
01 78 23.16	Maintenance Data
01 78 23.19	Preventative Maintenance Instructions
01 78 29	Final Site Survey
01 78 33	Bonds
01 78 36	Warranties
01 78 39	Project Record Documents
01 78 43	Spare Parts
01 78 46	Extra Stock Materials
01 78 53	Sustainable Design Closeout Documentation

contractor has successfully completed all testing and training and submitted the required commissioning and closeout submittals. Functional completion may or may not be a contractual milestone. If functional completion is a contractual milestone, then the contractor needs to understand exactly how it is defined in the owner-contractor agreement and make sure that its affected subcontracts also incorporate the functional milestone requirements, including any penalties or bonuses associated with achieving function completion.

Functional completion will be a prerequisite to final completion but may not necessarily be a prerequisite to substantial completion. Substantial completion is defined in Subparagraph 2.3.17 of AGC Document No. 200 entitled *Standard Form of Agreement and General Conditions between Owner and Contractor* as follows:

> Substantial Completion of the Work ... occurs on the date when the Work is sufficiently complete in accordance with the Contract Documents so that the Owner may occupy or utilize the Project ... for the use for which it is intended.

Usually, the prerequisite for substantial completion is the receipt of the certificate of occupancy from the authority having jurisdiction that allows the owner to move in, occupy, and use the building. The operation of those systems that make the building habitable, such as the HVAC and plumbing systems, as well as those associated with the health and safety of building occupants, such as the life-safety and emergency power systems, need to be fully functional to receive the building certificate of occupancy and will be part of the commissioning process. However, other functional completion requirements that do not affect the owner's use of the building, such as the submission of O&M manuals and operating personnel training, should not be a prerequisite for substantial completion.

9.15 WORKING WITH AN OUTSIDE COMMISSIONING AUTHORITY

9.15.1 Contractor Interaction with Outside Commissioning Authority

The most common approach to commissioning a green building is for the owner to retain an outside independent commissioning authority. The owner-contractor agreement will usually require that the contractor work with the commissioning authority to ensure that all equipment and systems operate as required. The contractor must thoroughly understand its role in the commissioning process, required equipment and systems testing, and required documentation that must be submitted before, during, and after the commissioning process. Commissioning can be a very time-consuming and expensive process, and the contractor needs to fully account for the required commissioning in its bid or proposal to the owner as well as in its schedule. Substantial completion of a green building project is often predicated on successful completion of the commissioning process, and failure to complete the prescribed commissioning process within the contract time is not an excusable delay.

9.15.2 Understand the Commissioning Requirements

The project commissioning requirements should be included in the owner's request for bids or proposals. These requirements may be included totally in the project specifications or they may be included both in the project specifications and in a separate draft commissioning plan. To understand both the commissioning scope and responsibilities, the contractor needs to thoroughly review its contract with the owner.

Ideally, the project specifications addressing commissioning will be fully coordinated, and the general specification commissioning requirements contained in Division 01 will point to equipment- and system-specific commissioning requirements in their respective specification sections. Similarly, each equipment- and product-specific specification section will point back to the general commissioning requirements contained in Division 01. Unfortunately, this is not always the case, because different individuals and firms and even different individuals and specialties within the same firm draft the various specification sections. Typically, the architect will draft the general requirements for the building commissioning and an engineer will draft the equipment- and system-specific commissioning requirements. The general and equipment- and system-specific commissioning requirements may or may not be coordinated, and the contractor needs to determine this not only to determine the scope of its own commissioning responsibilities but also to ensure that commissioning requirements are fully addressed in each affected subcontract scope of work.

As noted, in addition to or in conjunction with the project's specifications, the owner's commissioning authority will probably also prepare a draft commissioning plan that should be included as part of the bid documents. Ideally, the owner's commissioning authority will be involved in the design process and have an opportunity to not only review the project specifications regarding commissioning but also be involved in drafting those sections to make sure they are coordinated with the draft commissioning plan.

If the contractor believes that information and requirements regarding the commissioning process are missing or if there are conflicting requirements, the contractor should seek a written clarification or addendum before submitting its bid or proposal. As discussed earlier in this chapter, commissioning is a key element in the green building process, and commissioning should start during the planning process and finish at the end of the warranty period. Commissioning is usually a requirement for third-party certification of a green building.

9.15.3 Subcontractor Commissioning Requirements

Another important reason for the contractor to thoroughly understand the project commissioning requirements is so that it can ensure that subcontract scopes of work include the required commissioning requirements. This will ensure that the contractor receives complete bids or proposals from its subcontractors that will be involved in the commissioning process and that no gaps or overlaps will occur.

The contractor also needs to recognize that green building construction is an emerging market, and many specialty contractors may not be fully aware of commissioning requirements (as discussed in Chapter 6). It is in the contractor's best interest to ensure that all potential subcontractors understand the commissioning scope so that they provide a complete bid or proposal for their scope of work. In addition, by helping specialty contractors understand their role in the commissioning process and scope of work, the contractor has increased the number of qualified bids it will receive not only on this project but also on future green building projects.

9.15.4 Interaction with the Commissioning Authority

The contractor will need to work closely with the commissioning authority throughout the project. It is not simply a matter of implementing the commissioning plan during the construction phase of the project and providing feedback on the commissioning process in the prescribed format at the end of the process. Regular and ongoing interaction with the owner's commissioning authority will be required throughout construction, commissioning, and closeout. Therefore, the contractor and its affected subcontractors need to include the additional work and interaction in their project bid or proposal.

9.15.5 Finalizing the Commissioning Plan

Normally, the contractor will interact with the owner's commissioning authority throughout the construction process. This interaction will typically start with a meeting to review and finalize the commissioning plan a short time after the start of construction. During this meeting, the scope of the draft commissioning plan will be reviewed, and the commissioning schedule will be discussed. This meeting will also review submittal requirements; the development of procedures and checklists for inspection, testing, and startup of equipment; and system testing requirements. Testing and inspection documentation, O&M manual development, and owner operating personnel training may also be discussed in this meeting. After this meeting, the owner's commissioning authority will finalize the commissioning plan based on the contractor's input.

9.15.6 Ongoing Commissioning Authority Interaction

Following the finalization of the commissioning plan, the contractor will continue to interact with the commissioning authority in the course of fulfilling its commissioning responsibilities. This interaction will include ongoing planning and coordination meetings as well as meetings preparing for specific equipment and system commissioning activities that require coordination and possible witnessing of testing by outside entities, such as the fire marshal or other authority having jurisdiction.

9.15.7 Equipment and System Documentation and Submittals

In order for the commissioning authority to prepare or review equipment- or system-specific inspection, testing, and startup procedures and checklists, it must obtain equipment and system documentation from the contractor. Because most equipment and systems that will be commissioned are procured and installed by subcontractors, the contractor needs to obtain needed information from its subcontractors. The subcontractors in turn must obtain information from their subcontractors and suppliers. For example, TAB of the HVAC system is usually performed by the TAB contractor that is subcontracted to the sheet metal contractor, and the sheet metal contractor needs to obtain the needed TAB documentation from its TAB subcontractor and then pass that information on to the contractor, who in turn will pass it on to the owner's commissioning authority.

The contractor and its subcontractors should keep commissioning documentation requirements in mind when requesting shop drawings and other technical information from suppliers and subcontractors. The required product information, format of product information, and number of copies of product information should be requested from suppliers and subcontractors upfront to avoid multiple submittals and handling multiple copies of the same information. Most of the time, the information required for review by the design team to ensure compliance with the specifications will be the same information required by the commissioning authority to ensure proper installation and operation.

Equipment and system information typically required by the owner's commissioning authority includes the following:

- Manufacturer instructions for installation, preparation for startup, and operation
- Routine maintenance requirements and troubleshooting procedures
- Factory or third-party testing procedures, results, and certifications
- Equipment operating characteristics, such as motor torque-speed curves, transformer load and no-load losses, fan curves, and pump curves
- Equipment operating efficiencies and losses along with test results and guarantees
- Equipment and system warranties and guarantees
- Installation and checkout instructions shipped with the equipment
- Any field checkout procedures used by manufacturers' field technicians to verify proper equipment and system installation and operation

9.15.8 Create a Commissioning File

The contractor should create a commissioning file at the beginning of the project and maintain the file throughout the project in preparation for the submission of O&M manuals and other commissioning information at the end of the project. As noted earlier in this chapter, a lot of the information that will be included in the commissioning file will duplicate information contained in other project files, such as shop drawings. However, by maintaining a separate file for commissioning, it will serve as a reference source for the contractor throughout the project and will eliminate the time-consuming and tedious effort to create this file at the end of the project.

This file will probably need to be maintained in hard copy, but creating a separate electronic file using portable document format (PDF) will allow the contractor to post the documents on the project Web site for use by the owner, commissioning authority, design team, and subcontractors, which will reduce its time in responding to requests for documents and copying them. In addition, the electronic format can be linked to record drawings, which would allow the owner's operating personnel ready access to the information for operation, maintenance, troubleshooting, and repair activities after occupancy wherever they are if posted on the owner's intranet or Web site.

Table 9-8 provides a suggested organization for the contractor's commissioning file. The actual contents of the commissioning file will vary from project to project depending on the extent of the commissioning process and the closeout documentation required.

9.15.9 Preparation and Submission of O&M Manuals

Normally, the contractor will be required to document the commissioning process in an organized manner for the following reasons:

- Provide a detailed record of the commissioning process and demonstrate that it was carried out in accordance with the commissioning plan.
- Demonstrate that building systems perform and interact in accordance with the owner's requirements as documented in the planning stage of the project.
- Provide a benchmark for equipment and system performance over the life of the facility.

O&M documentation is usually required to be organized in loose-leaf notebooks with indices and tabs that make it easy for the owner's operating personnel to find the information they need. However, the contractor may be required to provide this information in electronic format either in lieu of or in addition to the hard-copy O&M submittal. In fact, with building information

Table 9-8

Suggested Commissioning File Organization
• Owner's Project Requirements (OPR)
• Designer's Basis of Design (BOD)
• Contractor's contract documents including:
▪ Owner-Contractor Agreement with General, Supplemental, and Special Conditions
▪ Plans
▪ Specifications
▪ Addenda
▪ Requests for Information (RFI) with responses
▪ Change Orders
• Commissioning Authority's commissioning plan
• Approved manufacturer product submittals
• Manufacturer inspection, testing, and startup procedures
• Manufacturer operation, maintenance, and troubleshooting procedures
• PITS checklists, issue logs, issue resolutions, and reports
• Building automation and control system design, operation, test procedure and test report
• HVAC air and hydronic TAB test procedure and test report
• FT procedures, issue logs, issue resolutions, and reports
• Record drawings
• Equipment and system warranties and guarantees
• Owner operating personnel training plan, program, and record of training
• Other documents

management (BIM) technology, the O&M information for individual pieces of equipment and systems could be tied to the electronic record drawings and accessed by clicking on the equipment or system of interest. This information could then be accessed by the owner's operating personnel anywhere in the facility during operation, maintenance, troubleshooting, or repair activities.

Typical information requirements for each piece of equipment or system includes the following:

- Owner's original requirements
- Design requirements that translated the owner's requirements into measurable system design and operating criteria
- Drawings and specifications for equipment and systems
- Equipment and system information, including catalog cuts and shop drawings, that was required to be submitted by the contractor for review and approval by the architect and engineers

- Installation, operation, and maintenance requirements for equipment and systems
- Record drawings depicting the actual physical installation, wiring and control diagrams, piping and instrumentation diagrams (PIDs), and other information unique to the installation
- Description of equipment operation, including sequence of operation
- Copies of condensed operating instruction prepared by the contractor and laminated and mounted on the equipment, which includes any or all of the following procedures:
  - Startup
  - Shutdown
  - Emergency operation
  - Other information
  - Testing procedures and test results

9.16 CASE STUDY

Gilbane Building Company

Concord Hospital, New Hampshire

At Concord Hospital in Concord, New Hampshire, the commitment to building a sustainable project was backed by hospital leadership from the very beginning—a key factor in the success of any LEED project. The hospital's East and North Wing additions are registered with the USGBC to achieve LEED Certification for New Construction (version 2.1) after completion this fall.

Concord's East and North Wing additions are part of a large-scale, long-term facilities plan for the hospital. Construction includes a new and expanded Emergency Department; new and expanded Intensive Care Unit; capacity for 60 additional private rooms to allow for the transformation of current semi-private rooms to become private; four state-of-the-art operating rooms; and a total of 166,960 square feet of new space and 25,335 square feet of renovated space.

Shepley, Bulfinch, Richardson & Abbott (SBRA) of Boston is the architectural firm responsible for designing the project to meet LEED standards. Gilbane Building Company's Manchester, New Hampshire office is managing the construction. Concord is one of the first hospitals in New England to seek LEED status. When it comes to LEED certification, the entire project team—from the

Figure 9-5 Photo Courtesy of Frank Giuliani, Photographer.

architect to the construction manager to the Owner—plays a role in obtaining the necessary credits.

The additions to Concord Hospital involve several green elements and best practices. The facility's exterior envelope includes several large windows and curtain-wall openings for natural lighting. Other elements include:

Regional materials and recycling. Part of the requirements for LEED is the use of regional materials (from within a 500-mile radius of the project), which reduces the amount of fuel needed to bring them to the site. In addition, 5 percent of the specified construction materials must contain recycled content. At Concord, the project team hopes to achieve an additional 20 percent of locally manufactured material.

Construction Waste Management. The project team is anticipating recycling at least 50 percent of the waste generated by construction, with the goal of increasing the amount to 75 percent.

Trade Contractor Education. Also essential is making sure that everyone is aware of the LEED goals. Gilbane educates all subcontractors about what the requirements are as part of the company's orientation process.

Green roof. The Concord project will feature a green roof, a garden roof system that is becoming highly desirable for healthcare facilities because of its positive healing effects and unique design elements. In addition to the healing aspects, the green roof works to avoid a potential heat sink on the roof, as well as mitigate rain runoff from the building.

Indoor Air Quality. The design process addresses controllability of building systems and thermal comfort, while the construction process controls the installation of low volatile organic compound (VOC) emitting materials for indoor air quality.

Figure 9-6 Photo Courtesy of Frank Giuliani, Photographer.

Lessons Learned

- Ensure the equipment vendor and/or trade contractor's contract or purchase order accounts for the startup of all applicable equipment, startup assistance, as well as all necessary training, documentation submittal, and operation and maintenance manuals as outlined within the Commissioning Plan.
- Make sure the warranty period for the equipment begins when the client takes responsibility/ownership of the system and/or equipment; extended warranties may need to be purchased.
- The project specifications will need to address the green building requirements as well as the method for documenting compliance. This documentation process occurs throughout the course of the project. Contractors that wait until the end of a project to start assembling the required green building closeout information could face delays due to missing information, nonresponsive vendors or subcontractors, etc. Ensure the certification approach is established at the beginning of the project and is included as part of all vendor and subcontractor contractual requirements.

9.17 REFERENCES

American Society for Quality, *Quality Management Systems – Fundamentals and Vocabulary*, ANSI/ISO/ASQ Q9000-2000, 2000.

The Associated General Contractors of America (AGC), *An Introduction to Total Quality Management*, Washington, D.C., 1992, p. 12.

Ledbetter, W. B., and James L. Burati, Jr., "On the Trail of Quality Costs," *The Construction Specifier*, May 1990, p. 127.

appendix A

Glossary of Green Terms and Abbreviations

Adaptability: Design strategy that allows for multiple future uses in a space as needs evolve and change. Adaptable design is considered a sustainable building strategy because it reduces the need to resort to major renovations or tearing down a structure to meet future needs. (Seattle 2007)

Adaptive Reuse: Renovation of a building or site to include elements that allow a particular use or uses to occupy a space that originally was intended for a different use. (Invista 2007)

Agricultural Waste: Materials left over from agricultural processes (e.g., wheat stalks, shell hulls). Some of these materials are finding new applications as building materials and finishes. Examples include structural sheathing and particleboard alternatives made from wheat, rye, and other grain stalks, and panels made from sunflower seed hulls. (Seattle 2007)

Airborne Particulates: Total suspended particulate matter found in the atmosphere as solid particles or liquid droplets. Chemical composition of particulates varies widely, depending on location and time of year. Sources of airborne particulates include dust, emissions from industrial processes, combustion products from the burning of wood and coal, combustion products associated with motor vehicle or nonroad engine exhausts, and reactions to gases in the atmosphere. (Invista 2007)

Air Pollutant: Any substance in air that could, in high enough concentration, harm humans, other animals, vegetation, or material. Pollutants may include almost any natural or artificial composition of matter capable of being airborne.

They may be in the form of solid particles, liquid droplets, gases, or a combination thereof. Generally, they fall into two main groups: (1) those emitted directly from identifiable sources, and (2) those produced in the air by interaction between two or more primary pollutants, or by reaction with normal atmospheric constituents, with or without photoactivation. Exclusive of pollen, fog, and dust, which are of natural origin, about 100 contaminants have been identified. Air pollutants are often grouped into categories for ease in classification; some of the categories are solids, sulfur compounds, volatile organic chemicals, particulate matter, nitrogen compounds, oxygen compounds, halogen compounds, radioactive compounds, and odors. (EPA 2007)

Air Quality Construction Management Plan: A systematic plan for addressing construction practices that can impact air quality during construction and continuing on to occupation. (Seattle 2007)

Alternative Energy: Energy from a source other than the conventional fossil-fuel sources of oil, natural gas, and coal (i.e., wind, running water, the sun). Also referred to as "alternative fuel." (Invista 2007)

Alternative Fuels: Substitutes for traditional liquid, oil-derived motor vehicle fuels like gasoline and diesel. Includes mixtures of alcohol-based fuels with gasoline, methanol, ethanol, compressed natural gas, and others. (EPA 2007)

ASHRAE: American Society of Heating, Refrigeration, and Air Conditioning Engineers. (Invista 2007)

ASTM: American Society for Testing and Materials.

Bake-out: Process by which a building is heated in an attempt to accelerate VOC emissions from furniture and materials. (Invista 2007)

Benefit/Cost Analysis: An economic method for assessing the benefits and costs of achieving alternative health-based standards at given levels of health protection. (Invista 2007)

Bicycle Storage: Covered and/or secured storage for building occupants commuting by bicycle. This amenity is considered a sustainable building technique in that it encourages human-powered transportation options. Some local governments offer subsidies or incentives to include bicycle storage in an existing or proposed building project. (Seattle 2007)

Biodegradable: Capable of decomposing under natural conditions. (EPA 2007)

Biodiversity: Refers to the variety and variability among living organisms and the ecological complexes in which they occur. Diversity can be defined as the number of different items and their relative frequencies. For biological diversity, these items are organized at many levels, ranging from complete ecosystems to the biochemical structures that are the molecular basis of heredity. Thus, the term encompasses different ecosystems, species, and genes. (EPA 2007)

Biological Contamination: Contamination of a building environment caused by bacteria, molds and their spores, pollen, viruses, and other biological materials.

It is often linked to poorly designed and maintained HVAC systems. People exposed to biologically contaminated environments may display allergic-type responses or physical symptoms, such as coughing, muscle aches, and respiratory congestion. (Invista 2007)

Biomass: All of the living material in a given area; often refers to vegetation. (EPA 2007)

Bioremediation: The cleanup of a contaminated site using biological methods (e.g., bacteria, fungi, plants). Organisms are used to either break down contaminants in soil or water, or accumulate the contaminants in their tissue for disposal. Many bioremediation techniques are substantially less costly than traditional remediation methods using heat, chemical, or mechanical means. (Seattle 2007)

Bioswale: A technology that uses plants and soil and/or compost to retain and cleanse runoff from a site, roadway, or other source. (Seattle 2007)

Blackwater: Water that contains animal, human, or food waste. (EPA 2007)

Brownfields: Abandoned, idled, or underused industrial and commercial facilities or sites where expansion or redevelopment is complicated by real or perceived environmental contamination. They can be in urban, suburban, or rural areas. The EPA's Brownfields initiative helps communities mitigate potential health risks and restore the economic viability of such areas or properties. (EPA 2007)

Building Envelope: The exterior surface of a building's construction—the walls, windows, floors, and roof. Also called building shell. (EPA 2007)

Building Flush-Out: See Flush-Out.

Building-Related Illness: Diagnosable illness whose cause and symptoms can be directly attributed to a specific pollutant source within a building (e.g., Legionnaire's disease, hypersensitivity, pneumonitis). (See sick building syndrome; biological contamination). (EPA 2007)

Carbon Dioxide Monitoring: A method for determining indoor air quality by using the concentration of carbon dioxide as an indicator. Although the level of CO_2 is a good general indicator of air quality, it is reliant on the presence of certain conditions and must be applied accordingly. (Seattle 2007)

Carbon Dioxide Sensor: Device for monitoring the amount of carbon dioxide in an air volume. (Seattle 2007)

Carbon Monoxide (CO): A colorless, odorless, poisonous gas produced by incomplete fossil-fuel combustion. (EPA 2007)

Carcinogen: Any substance that can cause or aggravate cancer. (EPA 2007)

Cellulose Insulation: Insulation alternative to glass fiber insulation. Cellulose insulation is most often a mixture of waste paper and fire retardant, and has thermal properties that are often superior to glass fiber. Glass fiber batt insulation often contains formaldehyde, which can adversely affect indoor air quality and human health, and the glass fibers are hazardous if inhaled and irritating to the skin and eyes. Specify cellulose insulation with high recycled content for maximum environmental benefit. (Seattle 2007)

Certified Lumber: General shorthand term for lumber that has been certified as sustainable harvest by an independent certification authority. See Forest Stewardship Council. (Seattle 2007)

Charrette: A meeting held early in the design phase of a project, in which the design team, contractors, end users, community stakeholders, and technical experts are brought together to develop goals, strategies, and ideas for maximizing the environmental performance of the project. Research and many projects' experience has indicated that early involvement of all interested parties increases the likelihood that sustainable building will be incorporated as a serious objective of the project, and reduces the soft costs sometimes associated with a green design project. (Seattle 2007)

Chlorofluorocarbons (CFCs): A family of inert, nontoxic, and easily liquefied chemicals used in refrigeration, air conditioning, packaging, insulation, or as solvents and aerosol propellants. Because CFCs are not destroyed in the lower atmosphere, they drift into the upper atmosphere, where their chlorine components destroy ozone. (EPA 2007)

Cistern: Small tank or storage facility used to store water for a home or farm; often used to store rainwater. (EPA 2007)

Commissioning (Building): The process of ensuring that installed systems function as specified, performed by a third-party Commissioning Authority. Elements to be commissioned are identified, installation is observed, sampling is conducted, test procedures are devised and executed, staff training is verified, and operations and maintenance manuals are reviewed. (Seattle 2007)

Conservation Easement: Easement restricting a landowner to land uses that are compatible with long-term conservation and environmental values. (EPA 2007)

Construction and Demolition Waste: Waste building materials, dredging materials, tree stumps, and rubble resulting from construction, remodeling, repair, and demolition of homes, commercial buildings, and other structures and pavements. May contain lead, asbestos, or other hazardous substances. (EPA 2007)

Construction Site Recycling: See Construction Waste Management. (Seattle 2007)

Construction Waste Management: General term for strategies employed during construction and demolition to reduce the amount of waste and maximize reuse and recycling. Construction waste management is a sustainable building strategy in that it reduces the disposal of valuable resources, provides materials for reuse and recycling, and can promote community industries. (Seattle 2007)

Cradle-to-Cradle: A term used in life-cycle analysis to describe a material or product that is recycled into a new product at the end of its defined life. (Invista 2007)

Cradle-to-Grave: A term used in life-cycle analysis to describe the entire life of a material or product up to the point of disposal. Also refers to a system that handles a product from creation through disposal. (EPA 2007)

Daylighting: Using natural light in an interior space to substitute for artificial light. Daylighting is considered a sustainable building strategy in that it can reduce reliance on artificial light (and reduce energy use in the process) and when well designed, contributes to occupant comfort and performance. (Seattle 2007)

Demand-side Waste Management: Prices whereby consumers use purchasing decisions to communicate to product manufacturers that they prefer environmentally sound products packaged with the least amount of waste, made from recycled or recyclable materials, and containing no hazardous substances. (EPA 2007)

Dioxin: Any of a family of compounds known chemically as dibenzo-p-dioxins. Concern about them arises from their potential toxicity as contaminants in commercial products. Tests on laboratory animals indicate that dioxins are one of the more toxic anthropogenic (man-made) compounds. (EPA 2007)

Disassembly: Taking apart an assembled product. Design for disassembly in buildings allows building components to be readily reused and recycled. (Seattle 2007)

Drought Tolerance: The capacity of a landscape plant to function well in drought conditions. (Seattle 2007)

Durability: A factor that affects the life-cycle performance of a material or assembly. All other factors being equal, the more durable item is environmentally preferable, because it means less frequent replacement. However, durability is rendered moot as a factor if the material is replaced for aesthetic reasons prior to it actually wearing out. (Seattle 2007)

Embodied Energy: The total amount of energy used to create a product, including energy expended in raw materials extraction, processing, manufacturing, and transportation. Embodied energy is often used as a rough measure of the environmental impact of a product. (Seattle 2007)

Energy Analysis: Analysis of the energy use of a structure. (Seattle 2007)

Energy Management System: A control system capable of monitoring environmental and system loads and adjusting HVAC operations accordingly in order to conserve energy while maintaining comfort. (EPA 2007)

Energy Modeling: Process to determine the energy use of a building based on software analysis. Also called building energy simulation. Common simulation software programs are DOE-2 and Energy Plus. (Seattle 2007)

Energy Star: Program administered by the EPA that evaluates products based on energy efficiency. (Seattle 2007)

Engineered Lumber/Wood: Composite wood products made from lumber, fiber or veneer, and glue. Engineered wood products can be environmentally preferable to dimensional lumber, because they allow the use of waste wood and small-diameter trees to produce structural building materials. Engineered wood products distribute the natural imperfections in wood fiber over the product, making them stronger than dimensional lumber. This allows for less material to be used in each piece, another environmental benefit. Potential environmental drawbacks with engineered wood include impacts on indoor environmental quality from offgassing of chemicals present in binders and glues, and air and water pollution related to production. (Seattle 2007)

Environmental Footprint: For an industrial setting, this is a company's environmental impact determined by the amount of depletable raw materials and nonrenewable resources it consumes to make its products, and the quantity of wastes and emissions that are generated in the process. Traditionally, for a company to grow, the footprint had to get larger. Today, finding ways to reduce the environmental footprint is a priority for leading companies. An environmental footprint can be determined for a building, city, or nation as well, and gives an indication of the sustainability of the unit. (Invista & Seattle 2007)

Environmental Impact Statement: A document required of federal agencies by the National Environmental Policy Act for major projects or legislative proposals significantly affecting the environment. A tool for decision making, it describes the positive and negative effects of the undertaking and cites alternative actions. (EPA 2007)

Environmental Tobacco Smoke: Mixture of smoke from the burning end of a cigarette, pipe, or cigar and smoke exhaled by the smoker. (EPA 2007)

EPA: Environmental Protection Agency

Erosion: The wearing away of land surface by wind or water, intensified by land-clearing practices related to farming, residential or industrial development, road building, or logging. (Seattle 2007)

Fluorocarbons (FCs): Any of a number of organic compounds analogous to hydrocarbons in which one or more hydrogen atoms are replaced by fluorine. Once used in the United States as a propellant for domestic aerosols, they are now found mainly in coolants and some industrial processes. FCs containing chlorine are called chlorofluorocarbons (CFCs). They are believed to be modifying the ozone layer in the stratosphere, thereby allowing more harmful solar radiation to reach the Earth's surface. (EPA 2007)

Flush-Out: A period after finish work and before occupation that allows the building's materials to cure and release volatile compounds and other toxins.

A building flush-out procedure is normally followed, with specified time periods, ventilation rate, and other criteria. (Seattle 2007)

Fly Ash: A fine, glass-powder recovered from the gases of burning coal during the production of electricity. These micron-sized earth elements consist primarily of silica, alumina, and iron. When mixed with lime and water, the fly ash forms a cementitious compound with properties very similar to that of portland cement. Because of this similarity, fly ash can be used to replace a portion of cement in the concrete, providing some distinct quality advantages. The concrete is denser, resulting in a tighter, smoother surface with less bleeding. Fly ash concrete offers a distinct architectural benefit with improved textural consistency and sharper detail. (Invista & Seattle 2007)

Footprint (Building): The area of a building formed by the perimeter of the foundation. Shrinking the footprint of a building allows for more open space and pervious surface on a site. (Seattle 2007)

Footprint (Environmental): See Environmental Footprint

Forest Stewardship Council (FSC): A third-party certification organization, evaluating the sustainability of forest products. FSC-certified wood products have met specific criteria in areas such as forest management, labor conditions, and fair trade. (Seattle 2007)

Formaldehyde: A colorless, pungent, and irritating gas, CH_2O, used chiefly as a disinfectant and preservative and in synthesizing other compounds like resins. (EPA 2007)

Fungus (Fungi): Molds, mildews, yeasts, mushrooms, and puffballs, a group of organisms lacking in chlorophyll (i.e., are not photosynthetic) and that are usually nonmobile, filamentous, and multicellular. Some grow in soil, whereas others attach themselves to decaying trees and other plants from whence they obtain nutrients. Some are pathogens; others stabilize sewage and digest composted waste. (EPA 2007)

Glazing: Translucent or transparent element of a window assembly. Glazing can have properties that increase the window's thermal performance, including Low-Emissivity coatings, multiple panes, thermally broken spacers, etc. (Seattle 2007)

Gray Water: Domestic wastewater composed of wash water from kitchen, bathroom, and laundry sinks, tubs, and washers. (EPA 2007)

Gray Water Reuse: A strategy for reducing wastewater outputs from a building, by diverting the gray water into productive uses such as subsurface irrigation, or on-site treatment and use for nonpotable functions such as toilet flushing. Gray water reuse is restricted in many jurisdictions; check with local health and building officials. (Seattle 2007)

Green Design: A design, usually architectural, conforming to environmentally sound principles of building, material, and energy use. A green building, for example, might use solar panels, skylights, and recycled building materials. (Invista 2007)

Green Label: A certification program by the Carpet and Rug Institute for carpet and adhesives meeting specified criteria for release of volatile compounds. (Seattle 2007)

Green Roof: Contained green space on, or integrated with, a building roof. Green roofs maintain living plants in a growing medium on top of a membrane and drainage system. Green roofs are considered a sustainable building strategy in that they have the capacity to reduce stormwater runoff from a site, modulate temperatures in and around the building, have thermal insulating properties, can provide habitat for wildlife and open space for humans, and offer other benefits. (Seattle 2007)

Ground Cover: Low-growing plants often grown to keep soil from eroding and to discourage weeds. (Seattle 2007)

Halon: Bromine-containing compounds with long atmospheric lifetimes whose breakdown in the stratosphere causes depletion of ozone. Halons are used in firefighting. (EPA 2007)

Heat Island Effect: A "dome" of elevated temperatures over an urban area caused by structural and pavement heat fluxes, and pollutant emissions. (EPA 2007)

Heavy Metals: Metallic elements with high atomic weights (e.g., mercury, chromium, cadmium, arsenic, and lead); can damage living things at low concentrations and tend to accumulate in the food chain. (EPA 2007)

High Efficiency: General term for technologies and processes that require less energy, water, or other inputs to operate. A goal in sustainable building is to achieve high efficiency in resource use when compared to conventional practice. Setting specific targets in efficiency for systems (e.g., using only EPA Energy Star–certified equipment, furnaces with an AFUE rating above 90 percent) and designs (e.g., watts per square foot targets for lighting) help put this general goal of efficiency into practice. (Seattle 2007)

High-Performance Glazing: Generic term for glazing materials with increased thermal efficiency. (Seattle 2007)

HVAC (Heating, Ventilation, and Air Conditioning): General term for the heating, ventilation, and air-conditioning system in a building. System efficiency and design impact the overall energy performance of a home and its indoor environmental quality. (Seattle 2007)

Hydrocarbons (HC): Chemical compounds that consist entirely of carbon and hydrogen. (EPA 2007)

Indigenous Planting: Landscaping strategy that uses native plants. Provided the natives are placed in the proper growing conditions, such plantings can have low or zero supplemental water needs. (Seattle 2007)

Indoor Air Pollution: Chemical, physical, or biological contaminants in indoor air. (EPA 2007)

Indoor Air Quality (IAQ): The ASHRAE defines acceptable indoor air quality as air in which there are no known contaminants at harmful concentrations as determined by cognizant authorities and with which 80 percent or more people exposed do not express dissatisfaction. (Invista 2007)

Infiltration: (1) The penetration of water through the ground surface into subsurface soil or the penetration of water from the soil into sewers or other pipes through defective joints, connections, or manhole walls. (2) The technique of applying large volumes of wastewater to land to penetrate the surface and percolate through the underlying soil. (EPA 2007)

Infiltration Rate: The quantity of water that can enter the soil in a specified time interval. (EPA 2007)

Inflow: Entry of extraneous rainwater into a sewer system from sources other than infiltration, such as basement drains, manholes, storm drains, and street washing. (EPA 2007)

Integrated Pest Management (IPM): A mixture of chemical and other non-pesticide methods to control pests. (EPA 2007)

Integrated Waste Management: The complementary use of a variety of practices to handle solid waste safely and effectively. Techniques include source reduction, recycling, composting, combustion, and landfilling. (Invista 2007)

Lead (Pb): A heavy metal that is hazardous to health if breathed or swallowed. Its use in gasoline, paints, and plumbing compounds has been sharply restricted or eliminated by federal laws and regulations. (EPA 2007)

LEED™: A self-assessing green building rating system developed by the U.S. Green Building Council. LEED™ stands for Leadership in Energy and Environmental Design, and evaluates a building from a systems perspective. By achieving points in different areas of environmental performance, a building achieves a level of certification under the system. (Seattle 2007)

Life Cycle (of a Product): All stages of a product's development, from extraction of fuel for power to production, marketing, use, and disposal. (EPA 2007)

Life-Cycle Analysis (LCA): The assessment of a product's full environmental costs, from raw material to final disposal, in terms of consumption of resources, energy, and waste. Life-cycle analysis is used as a tool for evaluating the relative performance of building materials, technologies, and systems. (Invista & Seattle 2007)

Life-Cycle Inventory (LCI): An accounting of the energy and waste associated with the creation of a new product through use and disposal. (Invista 2007)

Light Shelf: A horizontal shelf positioned (usually above eye level) to reflect daylight onto the ceiling and to shield direct flare from the sky. (Seattle 2007)

Linoleum: A resilient flooring product developed in the 1800s, manufactured from cork flour, linseed oil, oak dust, and jute. Linoleum's durability, renewable inputs, antistatic properties, and easy-to-clean surface often make it classified as a green building material. (Seattle 2007)

Local/Regional Materials: Building products manufactured and/or extracted within a defined radius of the building site. For example, the U.S. Green Building Council defines local materials as those that are manufactured, processed, and/or extracted within a 500-mile radius of the site. Use of regional materials is considered a sustainable building strategy because these materials require less transport, reducing transportation-related environmental impacts. Additionally, regional materials support local economies, supporting the community goal of sustainable building. (Seattle 2007)

Low-Emissivity (low-E) Windows: Window technology that lowers the amount of energy loss through windows by inhibiting the transmission of radiant heat while still allowing sufficient light to pass through. (EPA 2007)

Low Toxic: Generic term for products with lower levels of hazard than conventional products. Specific criteria need to be applied to this term to make it meaningful in the selection of sustainable building materials. (Seattle 2007)

Low VOC: Building materials and finishes that exhibit low levels of offgassing, the process by which Volatile Organic Compounds (VOCs) are released from the material, impacting health and comfort indoors and producing smog outdoors. Low (or zero) VOC is an attribute to look for in an environmentally preferable building material or finish. See Volatile Organic Compound (VOC) for more information. (Seattle 2007)

Manual: See Operations Manual.

Material Safety Data Sheets (MSDSs): A compilation of information required under the OSHA Communication Standard on the identity of hazardous chemicals, health and physical hazards, exposure limits, and precautions. Section 311 of the Superfund Amendments and Reauthorization Act (SARA) requires facilities to submit MSDSs under certain circumstances. (EPA 2007)

MDF (Medium-Density Fiberboard): A composite wood fiberboard, used for cabinetry and other interior applications. MDF containing urea formaldehyde can contribute to poor indoor air quality. (Seattle 2007)

Mercury: A metal that is an odorless silver liquid at room temperature, converting to an odorless, colorless gas when heated. Mercury readily combines with other elements and accumulates in the environment. Mercury is toxic and causes a range of neurological, organ, and developmental problems. Fluorescent lights and old thermostats are two building-related products that can contain significant amounts of mercury. Newer fluorescent lights are available with substantially reduced amounts of mercury. (Seattle 2007)

Methane: A colorless, nonpoisonous, flammable gas created by anaerobic decomposition of organic compounds. A major component of natural gas used in the home. Methane has also been found to be a potent greenhouse gas. Methane from landfills, livestock, and composting operations can be captured and used as a fuel source for alternative energy production. (EPA & Seattle 2007)

Mulch: A layer of material (wood chips, straw, leaves, etc.) placed around plants to hold moisture, prevent weed growth, and enrich or sterilize the soil. (EPA 2007)

Natural Ventilation: Ventilation design that uses existing air currents on a site and natural convection to move and distribute air through a structure or space. Strategies include placement and operability of windows and doors, thermal chimneys, landscape berms to direct airflow on a site, and operable skylights. (Seattle 2007)

Nonrenewable Energy: Energy derived from depletable fuels (i.e., oil, gas, coal) created through lengthy geological processes and existing in limited quantities on the earth. (Invista 2007)

Nonrenewable Resource: A resource that cannot be replaced in the environment (e.g., fossil fuels) because it forms at a rate far slower than its consumption. (Invista 2007)

Offgassing: Release of volatile chemicals from a product or assembly. Many chemicals released from materials impact indoor air quality and occupant health and comfort. Offgassing can be reduced by specifying materials that are low- or no-VOC and by avoiding certain chemicals (e.g., urea formaldehyde) entirely. Controlling indoor moisture and specifying prefinished materials can also reduce offgas potential. (Seattle 2007)

On-Site Stormwater Management: Building and landscape strategies to control and limit stormwater pollution and runoff. Usually an integrated package of strategies, elements can include vegetated roofs, compost-amended soils, pervious paving, tree planting, drainage swales, and more. (Seattle 2007)

Operations and Maintenance Manual (O&M Manual): Manual developed to assist building occupants in operating and maintaining a green building and its features. Many features' effectiveness can be reduced or eliminated by the actions (or inaction) of occupants and maintenance crews. An operations manual usually includes product and system information and warranties, contact information, and other information required for effective operations and maintenance. (Seattle 2007)

Organic Compound: Vast array of substances typically characterized as principally carbon and hydrogen, but that may also contain oxygen, nitrogen, and a variety of other elements as structural building blocks. (EPA 2007)

OSB (Oriented Strand Board): A high-strength, structural wood panel formed by binding wood strands with resin in opposing orientations. OSB is environmentally beneficial in that it uses small-dimension and waste wood for its fiber; however, resin type should be considered for human health impact and the production process monitored for air pollutant emissions. (Seattle 2007)

Overhangs: Architectural elements on roofs and above windows that function to protect the structure from the elements or to assist in daylighting and control of unwanted solar gain. Sizing of overhangs should consider their purpose, especially related to solar control. (Seattle 2007)

Pathogens: Microorganisms (e.g., bacteria, viruses, or parasites) that can cause disease in humans, animals, and plants. (EPA 2007)

Particulate Pollution: Pollution made up of small liquid or solid particles suspended in the atmosphere or water supply. (Invista 2007)

Particulate: (1) Fine dust or particles (e.g., smoke). (2) Of or relating to minute discrete particles. (3) A particulate substance. (Invista 2007)

Passive Solar: Strategies for using the sun's energy to heat (or cool) a space, mass, or liquid. Passive solar strategies use no pumps or controls to function. A window, oriented for solar gain and coupled with massing for thermal storage (e.g., a Trombe wall), is an example of a passive solar technique. (Seattle 2007)

Phytoremediation: Low-cost option for site cleanup when the site has low levels of contamination that are widely dispersed. Phytoremediation (a subset of bioremediation) uses plants to break down or uptake contaminants. (Seattle 2007)

Pollution: Generally, the presence of a substance in the environment that, because of its chemical composition or quantity, prevents the functioning of natural processes and produces undesirable environmental and health effects. Under the Clean Water Act, for example, the term has been defined as the man-made or human-induced alteration of the physical, biological, chemical, and radiological integrity of water and other media. (Invista 2007)

Pollution Prevention: Techniques that eliminate waste before treatment, such as changing ingredients in a chemical reaction. Identifying areas, processes, and activities that create excessive waste products or pollutants in order to reduce or prevent them through alteration or elimination of a process. The EPA has initiated several voluntary programs in which industrial or commercial partners join with the EPA in promoting activities that conserve energy, conserve and protect the water supply, reduce emissions or find ways of utilizing them as energy resources, and reduce the waste stream. (Invista 2007)

Porous Paving: Paving surfaces designed to allow stormwater infiltration and reduce runoff. (Seattle 2007)

Post-Consumer Recycling: Use of materials generated from residential and consumer waste for new or similar purposes (e.g., converting wastepaper from offices into corrugated boxes or newsprint.) (EPA 2007)

Postconsumer Recycle Content: A product composition that contains some percentage of material that has been reclaimed from the same or another end use at the end of its former, useful life. (Invista 2007)

Postindustrial Material: Industrial manufacturing scrap or waste; also called preconsumer material. (Invista 2007)

Postindustrial Recycle Content: A product composition that contains some percentage of manufacturing waste material that has been reclaimed from a process generating the same or a similar product. Also called preconsumer recycle content. (Invista 2007)

Preconsumer Materials/Waste: Materials generated in manufacturing and converting processes such as manufacturing scrap and trimmings and cuttings. Includes print overruns, overissue publications, and obsolete inventories. (EPA 2007)

Public Transportation: Mass transit, including bus and light rail systems. Siting a building near public transit is considered a sustainable building strategy, because it facilitates commuting without the use of single-occupancy vehicles. (Seattle 2007)

Radon: A colorless, naturally occurring, radioactive, inert gas formed by radioactive decay of radium atoms in soil or rocks. Design strategies help reduce the amount of radon infiltration into a building and remove the gas that does infiltrate. (EPA 2007)

Rainwater Catchment/Harvest: On-site rainwater harvest and storage systems used to offset potable water needs for a building and/or landscape. Systems can take a variety of forms, but usually consist of a surface for collecting precipitation (roof or other impervious surface) and a storage system. Depending on the end use, a variety of filtration and purification systems may also be employed. (Seattle 2007)

Reclamation: Restoration of materials found in the waste stream to a beneficial use that may be other than the original use. (Invista 2007)

Recycled Content: The content in a material or product derived from recycled materials versus virgin materials. Recycled content can be materials from recycling programs (postconsumer) or waste materials from the production process or an industrial/agricultural source (preconsumer or postindustrial). (Seattle 2007)

Recycling: Process by which materials that would otherwise become solid waste are collected, separated or processed, and returned to the economic mainstream to be reused in the form of raw materials or finished goods. (Invista 2007)

Recycling Areas: Space dedicated to recycling activities is essential to a successful recycling program, both on the construction site and in the building after occupation. (Seattle 2007)

Recycling Bins: Containers to temporarily hold recyclable materials until they are transferred to a larger holding facility for pickup by a recycling service. Conveniently located bins increase recycling rates by allowing occupants to recycle more easily. Designing space for recycling bins is a physical reminder of a commitment to recycling. (Seattle 2007)

Refurbished: Products that have been upgraded to be returned to active use in their original form. Refurbishing is considered a form of reuse, and is preferable to recycling because it requires less processing and inputs to return a product to useful service. (Seattle 2007)

Regional Manufacture: Goods produced within a certain radius of the project site. Using regionally produced goods is considered a sustainable building strategy in that it reduces the transportation impacts associated with the product, it often allows for a better understanding of the production process and increases the likelihood that the product was manufactured in accordance with environmental laws, and it supports regional economies. (Seattle 2007)

Relite: Windows or translucent panels above doors or high in a partition wall intended to allow natural light to penetrate deeper into a building. (Seattle 2007)

Renewable Resources: A resource that can be replenished at a rate equal to or greater than its rate of depletion (e.g., solar, wind, geothermal, and biomass resources). (Invista 2007)

Renovation: Upgrade of an existing building or space that maintains the original structure of a building. (Seattle 2007)

Resource Conservation: Practices that protect, preserve, or renew natural resources in a manner that will ensure their highest economic or social benefits. (Invista 2007)

Reuse: Using a product or component of municipal solid waste in its original form more than once (e.g., refilling a glass bottle that has been returned or using a coffee can to hold nuts and bolts). Reuse is a sustainable building strategy in that it reduces the strain on both renewable and nonrenewable resources, and when materials are reused on or near the site of salvage, they reduce transportation-related environmental impacts. (EPA & Seattle 2007)

Salvage: Building materials diverted from the waste stream intended for reuse. (Seattle 2007)

Sick Building Syndrome: Building whose occupants experience acute health and/or comfort effects that appear to be linked to time spent therein, but where no specific illness or cause can be identified. Complaints may be localized in a particular room or zone, or may be spread throughout the building. (EPA 2007)

Sisal: A durable natural fiber used as a floor covering, derived from leaves of the sisal plant. (Seattle 2007)

Source Reduction: The design, manufacture, purchase, or use of materials to reduce the amount or toxicity of waste in an effort to reduce pollution and conserve resources (i.e., reusing items, minimizing the use of products containing hazardous compounds, extending the useful life of a product and reducing unneeded packaging). (Seattle 2007)

Staging: The sequencing and physical positioning of building materials on a construction site. Sustainable building pays particular attention to staging in order to minimize the impact to the construction site and protect materials from damage. (Seattle 2007)

Stakeholder: Any organization, governmental entity, or individual that has a stake in or may be impacted by a given approach to environmental regulation, pollution prevention, energy conservation, etc. (EPA 2007)

Straw-Bale Construction: Alternative building method using bales of straw for wall systems. The method uses an agricultural waste product in place of diminishing dimensional lumber and achieves high insulation values. It is a building method most appropriate for regions with relatively little precipitation. (Seattle 2007)

Structural Insulated Panel (SIP): Manufactured panels consisting of a sandwich of polystyrene between two layers of engineered wood paneling. Can be used for walls, roof, or flooring, and results in a structure that is very resistant to air infiltration. (Seattle 2007)

Subsidies: Economic incentives to engage in an activity or purchase a product. Subsidies can work for or against environmental protection. Governments and utilities will sometimes offer subsidies for technologies that decrease energy or water use. (Seattle 2007)

Sulfur Dioxide (SO_2): A heavy, smelly gas that can be condensed into a clear liquid; used to make sulfuric acid, bleaching agents, preservatives, and refrigerants; a major source of air pollution in industrial areas. (Invista 2007)

Sunshades: Devices for blocking unwanted solar gain. (Seattle 2007)

SFI (Sustainable Forestry Initiative)

Tipping Fee: Charge for the unloading or dumping of waste at a recycling facility, composting facility, landfill, transfer station, or waste-to-energy facility. (Invista 2007)

Total Volatile Organic Compounds: The total mass, typically in milligrams per cubic meter, of the organic compounds collected in air. (Invista 2007)

Toxic: Capable of having an adverse effect on an organism; poisonous, harmful, or deadly. (Invista 2007)

Trombe Wall: Thermal storage system used in passive solar design. A high-mass wall that stores heat from solar gain during the day and slowly radiates the heat back into the living space at night. (Seattle 2007)

Truck Tire Wash-Down Area: A strategy for removing dirt and other contaminants from construction vehicles in order to prevent stormwater contamination related to transport of contaminants off-site on vehicle tires. A specified area is created for wash-down, with structural controls in place to prevent wash-down waters from entering the storm system or the larger environment. (Seattle 2007)

Urea-Formaldehyde Foam Insulation: A material once used to conserve energy by sealing crawl spaces, attics, etc. that is no longer used because emissions were found to be a health hazard. (Seattle 2007)

Volatile Organic Compound (VOC): Organic substances capable of entering the gas phase from either a liquid or solid form. (Invista 2007)

Walk-off Mat: Design strategy for reducing the amount of contaminants introduced into an interior space by providing grating or other material to remove contaminants from shoes. A significant portion of contaminants in a building are brought in this way, impacting indoor environmental quality. (Seattle 2007)

Wastewater: The spent or used water from a home, community, farm, or industry that contains dissolved or suspended matter. (EPA 2007)

Waste Management Plan: See Construction Waste Management.

Wetlands: An area that is saturated by surface or ground water with vegetation adapted for life under those soil conditions, such as swamps, bogs, fens, marshes, and estuaries. (EPA 2007)

Window Shading: Any device for reducing unwanted heat gain from a window. (Seattle 2007)

REFERENCES

City of Seattle (Seattle), Department of Planning and Development, *Green Building Glossary*, www.seattle.gov/dpd/GreenBuilding/OurProgram/Resources/Greenbuildingglossary, June 28, 2007.

Invista Antron (Invista), *Green Glossary*, www.antron.net/content/resources/green_glossary/ant06_04.shtml, June 28, 2007.

United States Environmental Protection Agency (EPA), *Terms of Environment: Glossary, Abbreviations and Acronyms*, www.epa.gov/OCEPAterms/gterms.html, June 28, 2007.

appendix B

References and Additional Information

GREEN BUILDING ORGANIZATIONS AND RATING SYSTEMS

U.S. Green Building Council, www.usgbc.org

U.S. Green Building Council, Leadership in Energy and Environmental Design (LEED™), *Green Building Rating System for New Construction and Major Renovations*, (LEED-NC), Version 2.2, October 2005, www.usgbc.org/DisplayPage.aspx?CMSPageID=220.

U.S. Green Building Council, Leadership in Energy and Environmental Design (LEED™), *New Construction and Major Renovation Reference Guide*, Second Edition, September 2006.

U.S. Green Building Council, Leadership in Energy and Environmental Design (LEED™), *Green Building Rating System for Core and Shell Development*, (LEED-CS), Pilot Version, September 2003, www.usgbc.org/DisplayPage.aspx?CMSPageID=295.

U.S. Green Building Council, Leadership in Energy and Environmental Design (LEED™), *Green Building Rating System for Core and Shell Development*, (LEED-CS), Pilot Version, September 2003, www.usgbc.org/DisplayPage.aspx?CMSPageID=295.

U.S. Green Building Council, Leadership in Energy and Environmental Design (LEED™), *Green Building Rating System for Comercial Interiors*, (LEED-CI), Version 2, November 2004, http://www.usgbc.org/DisplayPage.aspx?CMSPageID=145.

The Green Building Initiative, www.thegbi.org.

The Green Building Initiative, Green Globes™, www.thegbi.org/greenglobes/.

Japan Sustainable Building Consortium (JSBC), *Comprehensive Assessment System for Building Environmental Efficiency* (CASBEE), www.ibec.or.jp/CASBEE/english/index.htm.

National Resources Canada, *Green Building Assessment Tool*, (GBTool™), www.sbc.nrcan.gc.ca/software_and_tools/gbtool_e.asp.

THE ASSOCIATED GENERAL CONTRACTORS OF AMERICA

The Associated General Contractors of America (AGC), www.agc.org.

The Associated General Contractors of America, *The Contractors' Guide to BIM*, First Edition.

The Associated General Contractors of America, *Designing for Effective Sediment and Erosion Control*, 2001.

The Associated General Contractors of America, *Field Manual on Sediment and Erosion Control Best Practices*, 2002.

The Associated General Contractors of America, *Constructing an Environmental Management System: Guidelines and Templates for Contractors*, 2007.

The Associated General Contractors of America, *Construction Guidebook for Managing Environmental Exposures*, 2000.

The Associated General Contractors of America, *Project Delivery Systems for Construction*, Second Edition, 2004.

The Associated General Contractors of America, *An Introduction to Total Quality Management*, 1992, p. 12.

THE CONSTRUCTION SPECIFICATIONS INSTITUTE

The Construction Specifications Institute (CSI), www.csinet.org.

The Construction Specifications Institute, *MasterFormat™: Master List of Numbers and Titles for the Construction Industry*, 1995 Edition, 1996.

The Construction Specifications Institute, *MasterFormat™ Master List of Numbers and Titles for the Construction Industry*, 2004 Edition, 2004.

SHEET METAL AND AIR CONDITIONING CONTRACTORS' ASSOCIATION, INC.

Sheet Metal and Air Conditioning Contractors' Association, Inc. (SMACNA), www.smacna.org.

Sheet Metal and Air Conditioning Contractors' Association, Inc., *IAQ Guidelines for Occupied Buildings Under Construction*, First Edition, 1995.

Sheet Metal and Air Conditioning Contractors' Association, Inc., *Early Startup of Permanently Installed HVAC Systems*, SMACNA Position Paper, Undated.

GLOSSARIES OF TERMS, ABBREVIATIONS, AND ACRONYMS

American Society for Quality, *Quality Management Systems—Fundamentals and Vocabulary*, ANSI/ISO/ASQ Q9000-2000, 2000.

ASTM International, *Standard Terminology for Sustainability Relative to the Performance of Buildings*, ASTM Standard E 2114—06a, 2006.

City of Seattle (Seattle), Department of Planning and Development, *Green Building Glossary*, www.seattle.gov/dpd/GreenBuilding/OurProgram/Resources/Greenbuildingglossary, June 28, 2007.

Invista Antron (Invista), *Green Glossary*, www.antron.net/content/resources/green_glossary/ant06_04.shtml, June 28, 2007.

Lighting Design Lab (LDL), Northwest Energy Efficiency Alliance, *Lighting Glossary*, lightingdesignlab.com/library/glossary.htm, June 28, 2007.

United States Department of Energy, Energy Information Administration (EIA), *Glossary*, www.epa.gov/OCEPAterms/aaad.html, June 28, 2007.

United States Environmental Protection Agency (EPA), *Terms of Environment: Glossary, Abbreviations and Acronyms*, www.epa.gov/OCEPAterms/gterms.html, June 28, 2007.

Washington State University Cooperative Extension Energy Program (WSU), 2007.

SELECT MATERIAL AND EQUIPMENT STANDARDS AND LISTING

ASTM International, Standard Practice for Data Collection for Sustainability Assessment of Building Products, ASTM Standard E 2129 – 05, 2005.

Bay Area Air Quality Management District, BAAQMD Rules & Regulations, www.baaqmd.gov/dst/regulations/index.htm, August 21, 2007.

The Carpet and Rug Institute, Green Label/Green Label Plus, www.carpet-rug.com/commercial-customers/green-building-and-the-environment/green-label-plus/index.cfm, August 21, 2007.

Forest Stewardship Council, Principles and Criteria, www.fscus.org/standards_criteria/, August 21, 2007.

Froeschle, Lynn M., "Environmental Assessment and Specification of Green Materials," *The Construction Specifier*, October 1999, p. 53.

Green Seal, Green Seal Standards and Certification, www.greenseal.org/certification/standards.cfm, August 21, 2007.

South Coast Air Quality Management District, Rules and Regulations, www.aqmd.gov/rules/, August 21, 2007.

U.S. Environmental Protection Agency and the U.S. Department of Energy, Energy Star, www.energystar.gov/index.cfm?c=home.index, August 21, 2007.

SELECT GREEN BUILDING STUDIES

Fowler, K. M., and E. M. Rauch, *Sustainable Building Rating Systems Summary*, Pacific Northwest National Laboratory Operated by Battelle for the U.S. Department of Energy, Contract DE-AC05-76RL061830, July 2006.

Pulaski, Michael H., Editor-in-Chief, *Field Guide for Sustainable Construction*, Partnership for Achieving Construction Excellence/The Pennsylvania State University/Pentagon Rennovation and Construction Program Office, June 2004.

SELECT BUILDING COMMISIOING GUIDES AND SPECIFICATIONS

Portland Energy Conservation, Inc., *Model Commissioning Plan and Guide Specifications*, U.S. Department of Energy, Contracts DE-AP51-96 R020748 and DE-AP51-95 R020661, Version 2.05, February 1998, www.peci.org/library/mcpgs.htm.

U.S. Department of Energy, *Building Commissioning: The Key To Quality Assurance*, Rebuild America Guide Series, Undated, www.peci.org/library/PECI_BldgCxQA1_0500.pdf

SELECT GREEN BUILDING WEB SITES

The Associated General Contractors of America, *Environmental: Green Construction*, www.agc.org/environment

BuildingGreen.Com, www.buildinggreen.com

National Institute of Building Sciences, *The Whole Building Design Guide*, www.wbdg.org/design/sustainable.php

U.S. Environmental Protection Agency, *Green Buildings*, www.epa.gov/greenbuilding/

U.S. Environmental Protection Agency, *Energy Star*, www.energystar.gov/

The Construction Industry Compliance Assistance Center, *Green Building*, www.cicacenter.org/gb.html

SELECT BOOKS ON GREEN BUILDING DESIGN AND ESTIMATING

Kilbert, Charles J., *Sustainable Construction: Green Building Design & Construction*, John Wiley & Sons, Inc., 2005.

Means, R. S., *Green Building: Project Planning, and Cost Estimating*, John Wiley & Sons, Inc., 2002.

Mendler, Sandra, William Odell, and Mary Ann Lazarus, *The HOK Guidebook to Sustainable Design*, Second Edition, John Wiley & Sons, Inc., 2006.

Spiegel, Ross, and Dru Meadows, *Green Building Materials: A Guide to Product Selection and Specification*, Second Edition, John Wiley & Sons, Inc., 2006.

SELECT GREEN BUILDING PERIODICALS AND NEWSLETTERS

Building Design + Construction, Reed Business Information, www.bdcnetwork.com

eco-structure, www.eco-structure.com/index.cfm

Environmental Design + Construction, www.edcmag.com

GreenBuzz, GreenBiz.com, www.greenbiz.com/enewsletter

Greener Buildings, GreenBiz.com, www.greenerbuildings.com

GreenSource, McGraw-Hill, www.greensource.construction.com

iGreenBuild.com: The Voice of Sustainable Design and Construction, www.igreenbuild.com

Journal of Green Building, College Publishing, www.collegepublishing.us/journal.htm

Index

Environmental Benefits Statement

This book is printed with soy-based inks on presses with VOC levels that are lower than the standard for the printing industry. The paper, Rolland Enviro 100, is manufactured by Cascades Fine Paper Group and is made from 100 percent postconsumer, de-inked fiber, without chlorine. According to the manufacturer, the following resources were saved by using Rolland Enviro 100 for this book:

Mature trees	Waterborne waste not created	Waterflow saved (in gallons)	Atmospheric emissions eliminated	Energy not consumed	Natural gas saved by using biogas
225	103,500 lbs.	153,000	21,470 lbs.	259 million BTUs	37,170 cubic feet